Workshop Calculus

Guided Exploration with Review

Volume 1

Springer
New York
Berlin
Heidelberg
Barcelona
Budapest
Hong Kong
London
Milan
Paris
Santa Clara
Singapore
Tokyo

Workshop Calculus

Guided Exploration with Review

Volume 1

Nancy Baxter Hastings
Dickinson College

With contributing authors:

Priscilla Laws
Christa Fratto
Kevin Callahan
Mark Bottorff

Springer

Textbooks in Mathematical Sciences

COVER: Cover art by Kelly Alsedek at Dickinson College, Carlisle, Pennsylvania.

Library of Congress Cataloging-in-Publication Data
Baxter Hastings, Nancy.
Workshop calculus: guided exploration with review/Nancy Baxter Hastings.
p. cm. — (Textbooks in mathematical sciences)
Includes bibliographical references and index.
ISBN 0-387-94611-X (v. 1: softcover: alk. paper)
1. Calculus. I. Title. II. Series.
QA303.B3598 1996
515—dc20 95-47550

Printed on acid-free paper.

Production managed by Francine McNeill; manufacturing supervised by Jacqui Ashri.
Typeset by University Graphics, Inc., York, PA.
Printed and bound by Maple-Vail Book Manufacturing Group, Binghamton, NY.
Printed in the United States of America.

9 8 7 6 5 4 3 2 1

ISBN 0-387-94611-X Springer-Verlag New York Berlin Heidelberg SPIN 10517538

To my husband,
David,
and our family,
Erica, Benjamin, Karin and Mathew,
Mark and Margie, and John and Laura.

Preface

TO THE INSTRUCTOR

I hear, I forget.
I see, I remember.
I do, I understand.

Anonymous

OBJECTIVES OF WORKSHOP CALCULUS

1. Impel students to be active learners, rather than passive observers.
2. Help students to develop confidence about their ability to think about and do mathematics.
3. Encourage students to read, write, and discuss mathematical ideas.
4. Enhance students' understanding of the fundamental concepts underlying the calculus.
5. Prepare students to use calculus in other disciplines.
6. Inspire students to continue their study of mathematics.
7. Provide an environment where students enjoy learning and doing mathematics.

THE WORKSHOP APPROACH

Workshop Calculus: Guided Exploration with Review provides students with a gateway into the study of calculus. The program integrates a review of basic pre-calculus ideas with the study of concepts traditionally encountered in beginning calculus: functions, limits, derivatives, integrals, and an introduction to integration techniques. It seeks to help students develop the confidence, understanding, and skills necessary for using calculus in the natural and social sciences, and for continuing their study of mathematics.

In the workshop environment, lectures are replaced by an interactive teaching format where no formal distinction is made between classroom and laboratory work. Students learn by doing and by reflecting on what they have done. As the instructor, you respond to students as they learn, rather than students responding to you as you lecture.

In a typical Workshop Calculus class, new ideas and concepts are introduced in a brief, intuitive way: no formal definitions, no proofs of theorems, and no detailed examples—at least not yet. The purpose of this initial presentation is to help guide the students' thoughts in appropriate directions without giving anything away. Students then work collaboratively on activities in their Student Activity Guides, where they examine the behavior of mathematical systems in much the same way that science students explore natural phenomena. These tasks are designed to help students think like mathematicians—to make observations and connections, ask questions, explore, guess, learn from their errors, share ideas, read, write, and discuss mathematics—while working with their peers.

As students work on the assigned tasks, move from group to group, guiding discussions, posing questions, and responding to queries. If students are having difficulty, ask them to explain what they are trying to do, and then respond using their approach, trying not to fall into the let-me-show-you-how-to-do-this mode. Try to give only enough guidance to help them overcome their immediate problems, and try not to pick up a pencil or touch the keyboard.

After completing the activities, lead a class discussion, encouraging students to reflect on their own experiences. At this point, you can summarize what has been happening, present additional theoretical material, or give a minilecture. These presentations make sense to the students since they relate directly to the tasks they have been doing.

INSTRUCTIONAL MATERIALS

A key aspect of the Workshop Calculus materials is their flexibility. The length of class sessions, the balance between lecture and laboratory time, the type of technology that is used, the intended mathematical level, and the specific computer instructions can be varied by the local instructor. Although the two-volume set of materials is intended for a year-long integrated Pre-calculus/Calculus I course, subsets of the materials can be used in one-semester Pre-calculus and Calculus I courses, or they can be supplemented for use in an Advanced Placement Calculus course. Moreover, the Activity Guide may be used as a stand-alone book or in conjunction with other materials.

The *Workshop Calculus Student Activity Guide* is a collection of guided inquiry notes presented in a workbook format. Students make predictions, do calculations, and enter observations directly in the guide. As students begin to use the materials,

- encourage them to work together in groups of two to four.
- urge them to read carefully the small blurbs at the beginning of each section and prior to each task. These blurbs summarize what they have done and point the way to what is to come.
- ask them to write their responses to homework questions on separate sheets of paper, not in their Activity Guides.
- have them tear out the pages from their Activity Guide, for the section on which they are currently working, and place them in a three-ring binder. These pages can then be interspersed with lecture/discussion notes, answers to homework exercises, supplemental activities, and so on.

Workshop Calculus utilizes several software tools. Students use a motion detector to create distance versus time functions and to analyze their behavior. They use a computer algebra system (CAS) to do symbolic, numerical, and graphical manipulations, and they use the mathematical programming language ISETL to construct functions and to form mental images associated with abstract mathematical ideas, such as the process of a function or the limiting behavior of a function. Although the Workshop Calculus materials use particular software tools, there are no references in the text to a particular type of computer platform or to a specific CAS package.

Ideally, a microcomputer should be available for each group of two to four students. Each computer should be equipped with

- a Microcomputer-Based Laboratory (MBL) interface and software to support ultrasonic motion detection, available for both Macintosh and PC platforms from Vernier Software, 8565 S.W. Beaverton-Hillsdale Highway, Portland, OR 97225;
- a computer algebra system, such as Mathematica, Maple, or Derive; and
- the programming language ISETL, available for both Macintosh and PC platforms from West Publishing Company.

Model handouts for each of the software tools, including "Using the Motion Detector in Workshop Calculus," "Using ISETL in Workshop

Calculus," and "Using Mathematica in Workshop Calculus," are available electronically from Springer-Verlag. Each handout provides an overview of the features of the software tool that are used in Workshop Calculus. Although these handouts were developed for use at Dickinson College, where Workshop Calculus students use Mathematica on Macintosh computers, they can be easily customized for use at other institutions. Most of the information contained in the handouts is *not* contained in the Student Activity Guide, since it depends on the particular system and tools being used. Students, however, need this information to do the tasks. You should download the Dickinson College version of the handouts, make the necessary modifications for your locale, and distribute copies to your students.

In addition to the handouts on the various software tools, a set of notes to the instructor is available electronically from Springer-Verlag. These notes contain topics for discussion and review, solutions to homework exercises, and suggested timings for each task.

ACKNOWLEDGMENTS

The Workshop Calculus materials were developed in consultation with my physics colleague, Priscilla Laws, and one of my former students, Christa Fratto. Priscilla was the impelling force behind the project. She developed many of the applications that appear in the text, and her award-winning Workshop Physics project provided a model for the Workshop Calculus materials and the underlying pedagogical approach. She and Ron Thornton of Tufts University modified the MBL tools—which they had developed for use by physics students—for use in Workshop Calculus.

Christa Fratto, who graduated from Dickinson College in 1994, started working on the materials as a Dana Student Intern during the summer of 1992. She quickly became a partner in the project. She tested activities, offered in-depth editorial comments, developed problem sets, helped collect and analyze assessment data, and supervised the student assistants for the Workshop Calculus classes. Following graduation, Christa wrote the instructor notes that accompany the Student Activity Guide and the handouts for the software tools. She currently teaches at the Suffield Academy in Connecticut, where she is integrating the Workshop Calculus materials into her Pre-calculus and Advanced Placement Calculus classes.

Other major contributors include Kevin Callahan and Mark Bottorff, who helped design, write, and test initial versions of the material while on the faculty at Dickinson College. Kevin is now using the materials at California State University at Hayward, and Mark is pursuing his Ph.D. in Mathematical Physics.

A number of other colleagues have tested the materials and provided constructive feedback. Michael Kantor has been especially helpful in this regard. Based on his experience using the materials at both Guilford College and Knox College, he provided useful commentary on what worked and what did not, and suggested alternative ways to present ideas. Other colleagues who tested the materials and offered helpful comments include Peter Martin, Shari Prevost, Barry Tesman, Jack Stodghill, and Blayne Carroll at Dickinson College; Carol Harrison at Susquehanna University; Sandy Skidmore, Julia Clark, and Debra Etheridge at Emory and Henry College; and Nancy Johnson at Lake Brantley High School.

The Dickinson College students who assisted in Workshop Calculus courses helped make the materials more learner-centered and more user-friendly. These students include Jennifer Becker, Jason Cutshall, Amy Demski, Christa Fratto, Kimberly Kendall, Greta Kramer, Russell LaMantia, Susan Nouse, Alexandria Pefkaros, Benjamin Seward, Melissa Tan, and Jennifer Wysocki. Greta Kramer played a particularly active role. She succeeded Christa Fratto as the supervisor for the student assistants and helped her develop the answer keys that are contained in the instructor notes. In addition, Hannah Hazard, Linda Mellott, Matthew Parks, and Katherine Reynolds worked on the project as Dana Student Interns, reviewing the materials and analyzing assessment data. Sarah Buchan proofread the final version.

The development of the Workshop Calculus materials was also influenced by helpful suggestions from Jack Bookman of Duke University, Steve Davis of Davidson College, and Murray Kirch of the Richard Stockton College of New Jersey, who served as outside reviewers for the manuscript; Ed Dubinsky of Purdue University, who served as the project's mathematics education research consultant; and David Smith of Duke University, who served as the project's outside evaluator.

An important aspect of the development of the Workshop Calculus materials is the ongoing assessment activities. With the help of Jack Bookman, who served as the project's outside evaluation expert, we have analyzed student attitudes and learning gains, observed gender differences, collected retention data, and examined performance in subsequent classes. The information has provided the program with documented credibility and has been used to refine the materials for publication.

Workshop Calculus is part of the Workshop Mathematics Program, which also contains courses in statistics and quantitative reasoning developed by Allan Rossman. The program has been generously supported by grants from the Fund for Improvement of Post Secondary Education (FIPSE #P116B50675 and FIPSE #P116B11132), the National Science Foundation (NSF/USE #9152325, NSF/DUE #9450746, and NSF/DUE #9554684), and the Knight Foundation. Joanne Weissman manages the Workshop Mathematics Program. She keeps the program running smoothly and keeps us organized and on task.

Publication of the *Workshop Calculus Student Activity Guide* marks the culmination of five years of testing and development. We have enjoyed working with Jerry Lyons, Editorial Director of Physical Sciences at Springer-

Verlag. Jerry is a kindred spirit, who shares our excitement and understands our vision. We appreciate his support, value his advice, and enjoy his friendship. And, finally, we wish to thank Kim Banister, who did the illustrations for the manuscript. In her drawings, she caught the essence of the workshop approach: students exploring mathematical ideas, working together, and enjoying the learning experience.

Nancy Baxter Hastings
Mathias Professor of Mathematics
and Computer Science
Dickinson College

Preface

TO THE STUDENT

Everyone knows that if you want to do physics or engineering, you had better be good at mathematics. More and more people are finding out that if you want to work in certain areas of economics or biology, you had better brush up on your mathematics. Mathematics has penetrated sociology, psychology, medicine and linguistics . . . it has been infiltrating the field of history. Why is this so? What gives mathematics its power? What makes it work?

. . . the universe expresses itself naturally in the language of mathematics. The force of gravity diminishes as the second power of the distance; the planets go around the sun in ellipses, light travels in a straight line. . . . Mathematics in this view, has evolved precisely as a symbolic counterpart of this universe. It is no wonder then, that mathematics works: that is exactly its reason for existence. The universe has imposed mathematics upon humanity. . . .

Philip J. Davis and Rubin Hersh
Co-authors of *The Mathematical Experience*
Birkhauser, Boston, 1981

Task P-1: Why Are You Taking This Course?

Briefly summarize the reasons you decided to enroll in this calculus course. What do you hope you will gain by taking it?

Why Study Calculus?

Why should you study calculus? When students like yourself are asked their reasons for taking calculus courses, they often give reasons such as, "It's required for my major." "My parents want me to take it." "I like math." Mathematics teachers would love to have more students give idealistic answers such as, "Calculus is a great intellectual achievement which has made major contributions to the development of philosophy and science. Without an understanding of calculus and an appreciation of its inherent beauty, one cannot be considered an educated person."

Although most mathematicians and scientists believe that becoming an educated person ought to be the major reason why you should study calculus, we can think of two other equally important reasons for studying this branch of mathematics: (1) mastering calculus can provide you with conceptual tools that will contribute to your understanding of phenomena in many other fields of study and (2) the process of learning calculus can help you acquire invaluable critical thinking skills that will enrich the rest of your life.

What Is Calculus?

We have made several claims about calculus and its importance without saying anything about what it is or why it's so useful. It is difficult to explain this branch of mathematics to someone who has not had direct experience with it over an extended period of time. Hopefully, the overview that follows will help you get started with your study of calculus.

Basically, calculus is a branch of mathematics which has been developed to describe relationships between things which can change continuously. For example, consider the *mathematical relationship* between the diameter of a pizza and its area. You know from geometry that the area of a perfectly round pizza is related to its diameter by the equation

$$A = \tfrac{1}{4}\pi d^2.$$

You also know that the diameter can be changed *continuously*. Thus, you don't have to make just 9" pizzas or 12" pizzas. You could decide to make one that is 10.12" or one that is 10.13", or one whose diameter is halfway between these two sizes. A pizza maker could use calculus to figure out how the area of a pizza changes when the diameter changes a little more easily than a person who only knows geometry.

But it is not only pizza makers who could benefit by studying calculus. Someone working for the Federal Reserve might want to figure out how much metal would be saved if the size of a coin is reduced. A biologist might want to study how the growth rate of a bacterial colony in a circular petri dish changes over time. An astronomer might be curious about the accretion of material in Saturn's famous rings.

To a mathematician, all these questions can be answered by using calculus to find the relationship between the change in the diameter of a circle and its area. There is a beauty to the generality of this type of relationship. A mathematician might consider the fact that it is useful to so many other people purely academic.

Calculus and the Study of Motion

Ever since antiquity, philosophers and scientists have been fascinated with the nature of motion. Since motion seems intuitively to involve continuous change from moment to moment and seems to have regular patterns, it is not surprising that calculus can be used to describe motion in a very elegant manner.

Figure 1: Panathenaic Prize amphora depicting the motion of a runner, ca. 530 B.C. Attributed to Euphiletos Painter, Terracotta. The Metropolitan Museum of Art, all rights reserved.

The link between calculus and motion is quite fundamental. Isaac Newton, a seventeenth-century scientist, who made major contributions to the development of calculus, did so because he was primarily interested in describing motion mathematically. If motion is continuous, then it is possible to use calculus to find the speed of a runner or jumper at various times, if the relationship between his position and time is known. *Differential calculus* can be used to find the link between the change in position and velocity. Conversely, *integral calculus* can be used to find the position of a moving person whose velocity is known.

The potential of finding speed from changing positions and position from changing speeds can be illustrated by considering a sequence of photographs of a leaping boy taken by an eccentric artist-photographer, Eadweard Muybridge.[1]

Figure 2: Sequence of photos of a young boy leap frogging over the head of a companion. This series was taken by a famous artist-photographer Eadweard Muybridge. The original photos have been enhanced with lines and markers. *Boys Leapfrogging* (Plate 168). The Trout Gallery, Dickinson College. Gift of Samuel Moyerman, 87.4.8.

[1]Muybridge used a series of cameras linked electrically to obtain photos equally spaced in time. His mastery of technology was quite advanced for the time. Muybridge engaged in exploits other than photography, as he is reputed to have killed his wife's lover in a fit of rage. For this murder, a sympathetic jury acquitted him of his crime of passion. See G. Hendricks, *Eadweard Muybridge: The Father of the Motion Picture*, New York, Grossman, 1975.

Task P-2: Continuous Motion or Motion Containing Instantaneous Jumps and Jerks?

1. Suppose the time interval between frames in Figure 2 is 1/10th of a second. What might be happening to the position of the boy during the time between frames?

2. Do you think people and animals move from place to place in a series of little jumps and jerks or continuously? Explain the reasons for your answer.

3. Describe a way to use modern technology to determine if a leaping boy, like the one depicted in Figure 2, is moving in instantaneous jumps and jerks.

Summary

Most people believe that objects move *continuously* from position to position no matter how small the time interval between positions is. Thus, if the boy's big toe is at point A at one time and point B at another time, at any point between A and B there is a time when the boy's toe will be at that position.

As a result of the presumed continuity of motion, calculus is a powerful intellectual tool for exploring the nature of the boy's motion. For example, if you know the mathematical relationship between the boy's toe and time, you can use differential calculus to find its velocity—in other words, the rate at which the position of his toe changes. Conversely, if you know when the boy's toe is at a particular time and the mathematical relationship describing the velocity of the toe at each moment of time, you can use integral calculus to find a relationship for the position of the toe at each moment of time.

Calculus is not only useful in the study of motion. As we mentioned earlier, it is concerned with the study of mathematical relationships among two or more quantities that can vary continuously. The uses of calculus to study continuous changes are widespread and varied. It can be used to help understand many types of relationships, such as population changes of living organisms, the accumulation of the national debt, and the relationship between the concentration of chemicals and their reaction rates.

Using Computers and Collaboration to Study Calculus

As you complete the activities designed for this course, you will learn a lot about the nature of calculus. You will develop a conceptual understanding of the fundamental ideas underlying calculus: function, limit, derivative, antiderivative, and definite integral. You will discover how to use calculus to solve problems. Along the way, you will review the necessary algebraic and trigonometric concepts needed to study calculus.

The methods used to teach Workshop Calculus may be new to you. In the workshop environment, formal lectures will be replaced by an interactive teaching format. You will learn by doing and reflecting on what you have done. Initially, new ideas will be introduced to you in an informal and intuitive way. You will then work collaboratively with your classmates on the activities in this workbook—which will be referred to as your Student Activity Guide—exploring and discovering mathematical concepts on your own. You will be encouraged to share your observations during class discussions.

Although we can take responsibility for designing a good learning environment and attempting to teach you calculus, you must take responsibility for learning it. No one else can learn it for you. You should find the thinking skills and mathematical techniques acquired in this course useful in the future. Most importantly, we hope you enjoy the study of calculus and begin to appreciate its inherent beauty.

A number of the activities in this course will involve using computers to enhance your learning. You will be introduced to some computer tools that have been developed for recording scientific data, learning mathematical concepts, and doing calculations. These tools include the Microcomputer-Based Laboratory system, or MBL system, in which a motion detector, electronic interface, and software can be used to measure position and velocity changes over time, the mathematical programming language ISETL, and a computer algebra system. Using the computer will help you develop a conceptual understanding of important mathematical concepts and help you focus on significant ideas, rather than spending a lot of time on extraneous details.

Task P-3: Getting Started with Your Computer

Familiarize yourself with the computer which you will be using in this course.

1. Find out how to turn the machine on and off, and if necessary, how to access your account.
2. If your machine has a windows environment, practice using the mouse to point, click, highlight, and drag. Learn how to access the menus and manipulate the windows.
3. Find out how to name a new file, retrieve an old file, print a file, and discard a file.
4. If you will be saving information on a diskette, find out how to insert a disk in the machine, format it, save a file on the disk, and eject it from the machine.

Some Important Advice Before You Begin

- ***Work closely with the members of your group.*** Think about the tasks together. Discuss how you might respond to a given question. Share your thoughts and your ideas. Help one another. Talk mathematics.
- ***Read carefully the short blurbs at the beginning of each section and prior to each task.*** These blurbs summarize what you have done and point the way to what is to come. They contain important and useful information.
- ***Answer the questions in your Activity Guide in your own words.*** Work together, but when it comes time to write down the answer to a question, do not simply copy what one of your partners has written. Besides, submitting someone else's work as your own, even if you have discussed the ideas, may be viewed as plagiarism.

- ***Use separate sheets of paper for homework problems.*** Do not try to squish the answers in between the lines in your Activity Guide.

- ***Put together your own book.*** Remove the pages for the current section from your Activity Guide, and place them in a three-ring binder. Intersperse the pages with lecture and discussion notes, answers to homework problems, and handouts from your instructor.

- ***Think about what the computer is doing.*** Whenever you ask the computer to perform a task, think about how the computer might be processing the information that you have given it, keeping in mind:

  What you have commanded the computer to do.
  Why you asked it to do whatever it is doing.
  How it might be doing whatever you have told it to do.
  What the results mean.

- ***Switch typists on a regular basis.*** Do not become the designated typist or an ongoing observer for your group. Learning to use the computer is an important part of the learning process

- ***Have fun!***

Nancy Baxter Hastings
Mathias Professor of Mathematics
and Computer Science
Dickinson College

Contents

Volume 1

Name ______________________________ Section __________ Date ______

1

Unit 1:

FUNCTIONS

I always assumed I could learn the math, but I never knew I could figure it out for myself.

Workshop Calculus student
Dickinson College
Fall 1994

OBJECTIVES

1. Understand what a function is.
2. Model situations using functions.
3. Analyze the shape of the graph of a function.
4. Investigate the behavior of the tangent line to a smooth curve.

OVERVIEW

Functions are used to model real-life situations. For example, if you started a new business, you could use a function to model your earnings over time. You could then analyze your earnings by examining the function, asking questions such as: During what time intervals were your earnings increasing? During what time intervals were they decreasing? At what times did your earnings level off? If your earnings were growing and then they leveled off, did they start to grow again or decrease? Conversely, if your earnings were declining and they leveled off, did they fall some more or did they start to grow? At what times did your earnings reach a peak? A dip? In other words, at what times were your earnings maximized? Minimized? In addition to investigating the growth and decline of your earnings, you could also analyze the rate of change in your earnings. For instance, during periods of growth, how fast were your earnings increasing? Were they growing at a steady rate? An increasing rate? A decreasing rate? At what time was your rate of growth the largest? Was there a point when your rate of growth changed from increasing to decreasing, or vice versa? You could also ask similar questions about periods when your earnings were declining. Knowing the answers to these types of questions is obviously very important if you want to have a good handle on what is happening with your business. One way to do this is to model your earnings with a function and then use calculus to analyze its behavior. This is exactly what you will learn how to do in this course.

The primary objectives of this unit are to help you understand what a function is and how it can be used to model a situation, and to make a first pass at analyzing a function's behavior. To help you do this, you will use a motion detector to create distance versus time graphs that model your movements. You will observe how your movements affect the shape of the associated graph. In the process, you will develop a conceptual understanding of what it means for a function to be increasing, decreasing, concave up, or concave down. You will also develop an intuitive understanding of what a tangent line to a smooth curve is and how it behaves as it travels along the curve. You will investigate the relationship between the tangent line and the rate of change of the function.

The slope of the tangent line—or rate of change—turns out to be a key idea in this discussion. The only problem is that unless the underlying function is linear, you do not know how to find the value of the slope using algebraic methods. You can approximate the value, but you cannot find the exact value. You need something stronger. Calculus!

SECTION 1

Modeling Situations

A function is a process for finding the relationship between two entities such as position and time. Consider, for example, the sequence of photographs in Figure 1 depicting the motion of a horse. How can you use a function to describe the motion of the horse during this short time period? That is, how can you use a function to model this situation?

Although data are not given for this sequence, suppose Muybridge's as-

Figure 1: Selections from a photo sequence by Eadweard Muybridge. The lines near the horse's head (which have been enhanced for clarity) indicate the marking pole used by the photographer. The other lines indicate the breaks between successive frames. *Sulky Horse Walking* (Plate 588). The Trout Gallery, Dickinson College. Gift of Samuel Moyerman, 87.4.65.

sistant started a stopwatch the moment the first photograph was taken and that the time interval between each frame is 1/5th or 0.20 seconds—that is, assume frame 1 was taken at 0.00 seconds, frame 2 at 0.20 seconds, and so on. In addition, assume the length of the horse's head is 60 centimeters (cm). In the next task, you will use these assumptions to calculate the real distance from the tip of the horse's nose to the marker pole at each of the four times. You will then describe—or represent—this information which defines a function in several ways.

Task 1-1: Relating Position and Time

1. One of the objectives of this task is to calculate the "real" distance from the horse's nose to the marker pole in each frame. You can use a centimeter ruler to measure the distances in the photograph, but then you will need to convert these results to actual centimeters. To do this, you need to know the *scale* of Muybridge's photographs—for example, the scale for a given map might be that 1/4" (on the map) = 1 "real" mile. Use the assumption that the actual length of the horse's head is 60 centimeters to calculate the scale for the photographs.

   a. 1 cm (on the photograph) = ________ "real" centimeters

   b. 1/10 cm (on the photograph) = ________ "real" centimeters

2. Determine the real distance from the tip of the horse's nose to the marker pole in each of the four frames.

   a. Frame 1: $t = 0.00$ seconds

   b. Frame 2: $t = 0.20$ seconds

c. Frame 3: $t = 0.40$ seconds

d. Frame 4: $t = 0.60$ seconds

3. In one or two sentences, describe the relationship between the elapsed time on the stopwatch and the location of the nose relative to the marker pole to someone who has not seen the photographs.

4. Figure 1 represents the distance of the horse's nose from the marker pole pictorially. There are a number of different ways to represent the corresponding numerical data—that is, to display the actual distance at each of the four times. Find at least two different ways to do this. Make your representations as complete and accurate as possible.

a. Representation #1:

b. Representation #2:

Although there are only four frames in the sequence in Figure 1, you can estimate the location of the horse's nose at times other than those in which the photographs were made. The goal of the next task is to try to find a *process* that you can use to find the value of the position at any time between 0.00 and 0.60 seconds, not just the times represented by the four frames.

Task 1-2: Describing a Process for Finding the Position at a Given Time

1. What assumptions do you have to make about the nature of the motion of the horse's nose in order to make a reasonable estimate of its position (distance from the marker pole) at a time which is not a time when a photograph was made?

2. Is it possible for the horse's nose to be at two different positions at a single time?

3. Estimate the position of the horse's nose at 0.12 seconds and at 0.38 seconds.

a. Enter your estimates in the table below.

Time (seconds)	Position (centimeters)
0.12	
0.38	

b. In one or two sesntences, describe the process that you used to obtain your estimates in part a.

4. Suppose you wanted to estimate the position of the horse's nose at any time between 0.00 and 0.60 seconds. Describe the process you would use to obtain this estimation.

At any given time between 0 and 0.6 seconds, the horse's nose is at one and only one position (with respect to the marker pole). Moreover, you can describe a process for determining the position corresponding to a particular time. Consequently, you say that there is a *functional relationship* between position and time or that position is a *function of* time. The important observation is that corresponding to each input (time), there is exactly one output (position).

Apply the idea of a function to another situation. Suppose you have collected \$10 to buy pizza for your hungry friends. An 8" pizza costs \$5 and a 12" pizza costs \$10. Would it be better to buy two 8" pizzas for the gang or one 12" pizza? Is area a function of diameter in the mathematical sense? In the following task, you will find a function that *models* this situation and then use the function to determine which combination gives you more for your money.

Task 1-3: Using the Concept of Function to Buy Pizza

1. Assuming you have collected \$10 to buy pizza and that an 8" pizza costs \$5 and a 12" pizza costs \$10, offhand what do you think will give you the most pizza to eat? One 12" pizza or two 8" pizzas?

2. Use a formula to represent the functional relationship between the diameter d of a circle and its area A.

3. Give a verbal description of a process you can use to find the area of a pizza if you know its diameter.

4. In a functional relationship there is exactly one output corresponding to each input.

   a. Give the input quantity in the model for the pizza dilemma.

   b. Give the corresponding output quantity.

   c. Explain why there can be only one output for each given input. In other words, explain why the area formula for a pizza represents a function.

5. Now that you have used a function to model this situation, what pizza combination would you buy for your friends? One 12" pizza or two 8" pizzas? Use your function to justify your decision.

Unit 1 Homework After Section 1

- Complete the tasks in Section One in the Activity Guide. Be prepared to discuss them in class.
- As a general rule, do your homework on separate sheets of paper and place the sheets in your activity guide at the end of the appropriate section. Please don't try to scrunch your answers in with the description of the homework problem!
- There are a few basic mathematical ideas that you need to recall before you go on to the next section. Refresh your memories as you do HW1.1. Work together with the members of your group. Answer each other's questions. Share your results. Go through the problems quickly. If you need special help, see your instructor.

HW1.1 Review Exercise.

1. Lines and slopes. First recall a few definitions.

If $P(x_1,y_1)$ and $Q(x_2,y_2)$ are any two points on a given line, then the *slope* m of the line through P and Q is given by

$$m = \frac{y_2 - y_1}{x_2 - x_1}.$$

If m is the *slope* of a given line and b is the y-intercept, then the *slope-intercept equation* for the line is given by $y = m(x - x_1) + y_1$.

a. Find the slope of the line through the following pairs of points.
(1) $P(1,2)$ and $Q(3,6)$
(2) $R(0.5,1.6)$ and $T(-2,1.6)$
(3) $M(-1,-3)$ and $P(-4,5)$

b. Find the equation of each of the following lines.
(1) The line through the points $P(1,2)$ and $Q(3,6)$
(2) The line through the point $S(\frac{1}{2},\frac{1}{2})$ with slope $\frac{1}{2}$
(3) The line with slope -3 and y-intercept -10.67
(4) The line parallel to the horizontal axis that passes through the point $P(-4,5)$
(5) The line through the point $K(0,-3.4)$ that is parallel to the line $y = 9.8x + 6$

(6)

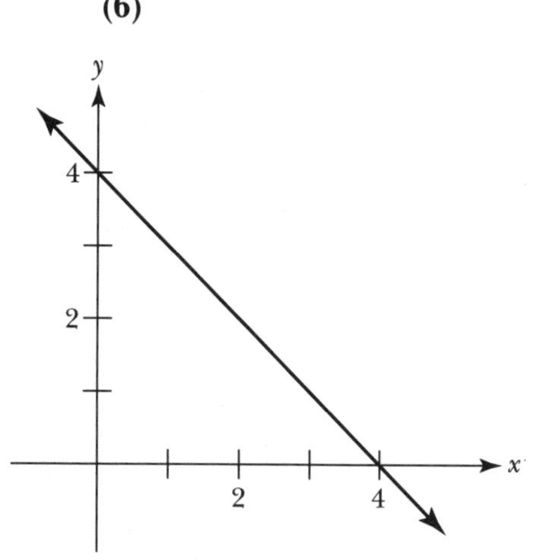

(7)

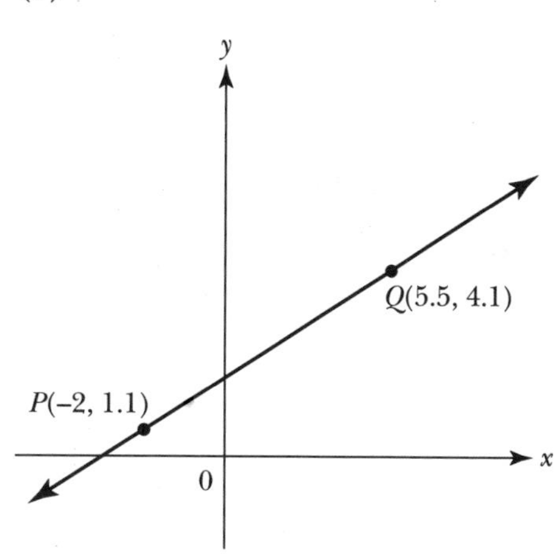

c. Sketch a graph of the following functions.

(1) $y = x$

(2) $y = -x$

(3) $x + y = 2$

(4) $y = 0$

(5) $y = \begin{cases} 2, & \text{if } x \leq 3 \\ x - 1, & \text{if } x > 3 \end{cases}$

2. Ways of denoting an interval.

Recall that there are several ways to denote an interval of real numbers. One way is to use a *number line,* where

denotes the set of all real numbers between a and b, but not including either a or b. This set can also be denoted using *open interval notation* (a,b) or using the *inequality* $a < x < b$, where x is any value between a and b. If you want to include a and/or b in the interval, draw a closed circle instead of an open circle on the number line; use a square bracket [and/or] instead of an open bracket with the interval notation; or use a less than or equal to sign $\leq$ in the inequality. Intervals that include either a or b, but not both, are called *half-open intervals.* Intervals that include both a and b are said to be *closed intervals.*

Practice converting between the ways. Each of the following intervals is represented using one of the three ways—that is, using a number line, interval notation, or an inequality. Express each interval two other ways.

a. $-6 \leq x \leq 3$

b. $0 < x \leq 2.5$

c. $x < -5$ or $x \geq 16$

d. $[-7,8)$

e. $(-2.1,6)$ or $(10,12]$

f. $(-\infty,0]$ or $(1.5,\infty)$

g.

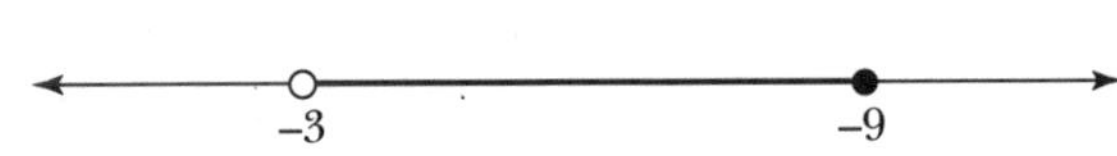

h.

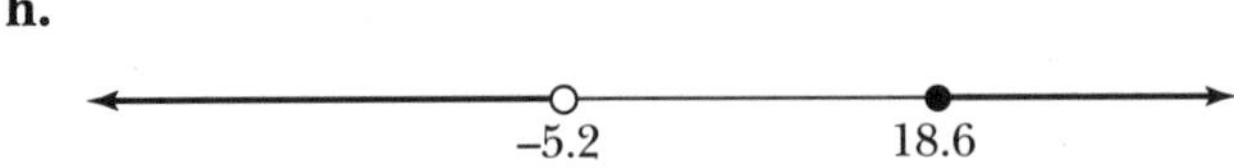

i.

3. A few important facts.

Briefly explain or show why each of the following statements makes sense to you.

a. You cannot divide a real number by 0 (and get a real number).

b. You cannot find the square root of a negative number (and get a real number).

c. The slope of a horizontal line is always 0.

d. The slope of a line which rises from left to right is always positive.

e. The slope of a line which falls from left to right is always negative.

f. The slope of a vertical line is undefined.

SECTION 2

Analyzing Linear Functions

You will begin your study of functions by using a motion detector to create your own functions. As you walk, jump, or run in front of the motion detector, it sends out a series of ultrasonic pulses which bounce off an object in its path (in this case, you) and back to the detector. The computer or calculator records the time it takes each sound pulse to make the round-trip between the detector and the object, uses this information to calculate the object's distance from the detector, and instantaneously displays the distance versus time graph generated by the motion on its screen.

In the first task[1] you will use the motion detector to create some *linear* functions—that is, functions whose graphs are straight lines—and then find an expression that models your curve. You will create increasing and decreasing functions and examine how varying the rate at which you move affects the shape of the graph representing your motion. You will examine some ways of representing a function and think about the meaning of domain and range.

Task 1-4: Creating Linear Functions

As you go through this task, read each of the questions very carefully before you begin. Try to do exactly what it says. Work as a team. Share the calculations. Compare your results. Talk about what is going on . Help each other out. Make sure each person takes a turn using the motion detector. You may try a number of times. Get the times right. Get the distances right.

Open the motion detector software.

1. Increase your distance slowly. Starting at the $\frac{1}{2}$-meter mark, increase your distance from the detector walking at a slow, constant pace. Sketch a graph of your function on the axes given below. Label your axes.

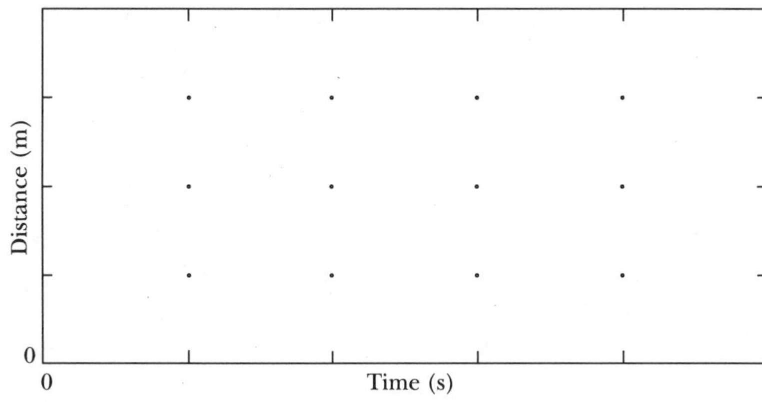

a. Use your graph to estimate your distance when

(1) $t = 1$ second

[1]The motion detector investigations are based on activities developed by Ron Thornton of Tufts University, David Sokoloff of the University of Oregon, and Priscilla Laws of Dickinson College.

(2) $t = 3.5$ seconds

b. In one or two sentences, describe the process you used in part a. That is, describe the process you used to find your distance at a given time by examining your graph.

c. Although you tried to walk at a steady pace, the graph representing your motion is probably bumpy. Try to find a line that approximates or "fits" your graph.

(1) Approximate the slope of the line that best fits your graph.

Note: In this case, the unit for the slope is meters per second.

(2) Find an equation for the line that best fits your graph, using the *point-slope equation* for the line and the value for the slope from part (1).

d. Before continuing on, save your graph in Data B. In the next part of this task, you will be asked to create a new graph and compare it to this one.

2. Go faster. Starting at the $\frac{1}{2}$-meter mark, increase your distance from the detector walking at a medium fast, constant pace. If you walk out of the range of the detector, adjust the upper bound on the horizontal axis.

Print a copy of the two graphs and paste it in your Activity Guide in the space below.

a. Find an expression which represents your new function by fitting a line to your graph.

b. Compare the two graphs which you created by increasing your distance moving at a constant pace.

(1) The slope represents your *rate of change* or your *velocity*. Describe how the rate at which you walk affects the value of the slope of the function modeling your motion.

(2) When you increase your distance from the detector, the slope of the associated graph is always positive. Explain why.

c. If s is the name of your function, then $s(t)$—which is read "s of t"—gives the output, or your distance from the detector, at time t. Use your graph to approximate the values of

(1) $s(0.5)$

(2) $s(1.5)$

d. Represent your new function by a *table* by approximating the values for $s(t)$ at the specified times.

t	0.5	1.0	1.5	2.0
$s(t)$				

e. Describe how the columns in the table relate to points on your graph.

f. The *domain* of a function is the set of all acceptable input values. In the case of a motion-detector generated function, the domain is the time interval during which the detector is collecting data.

Consider the new function you created above.

(1) Find a point in the domain of the function. State the value of the point below and mark the point on the horizontal axis of your graph.

(2) Give the value of a point that is not in the domain of your function.

(3) Find the domain of your new function.

(4) Describe in general how you can determine the domain of a function by looking at its graph.

g. Before continuing on, clear the data for both graphs. In the next two parts you will create two new graphs, but this time you will decrease your distance from the detector.

3. Decrease your distance slowly. Starting at the 3-meter mark, decrease your distance from the detector walking at a slow, constant pace. Sketch a graph of your function on the axes given below. Label your axes.

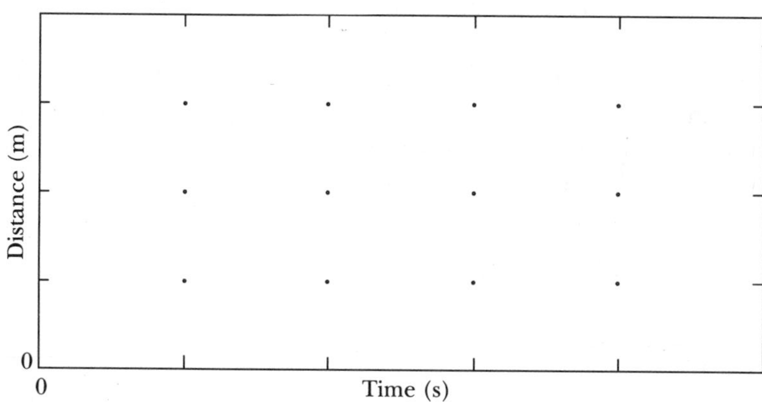

a. Find an expression which represents your function by fitting a line to your graph.

b. Use your expression to approximate your distance from the detector when t is 1 second.

c. Find the domain of your function.

d. The *range* of a function is the set containing all the function's possible outputs.

(1) Find the range of the function you created above.

(2) Describe how you can determine the range of a function by looking at its graph.

e. Before continuing on, save your graph in Data B, so you will be able to compare it to your next graph.

4. Create another graph, but this time walk more rapidly. Starting at the 3-meter mark, decrease your distance from the detector walking at a medium fast, constant pace. Print a copy of the two graphs and paste it in your Activity Guide in the space below.

a. Find the slope of the line that best fits the graph of your new function.

b. Compare the two graphs which you created by decreasing your distance moving at a constant pace.

(1) Describe how the rate at which you walk affects the value of the slope of the function modeling your motion.

(2) When you decrease your distance from the detector, the slope of the associated graph, and hence your rate of change or velocity, is always negative. Explain why.

c. Find the range of your new function.

d. Clear the data for both graphs.

5. Investigate what happens if you stand still in front of the detector—that is, if you neither increase nor decrease your distance.

a. Stand 1.5-meters from the detector for 5 seconds. Sketch the associated distance versus time graph.

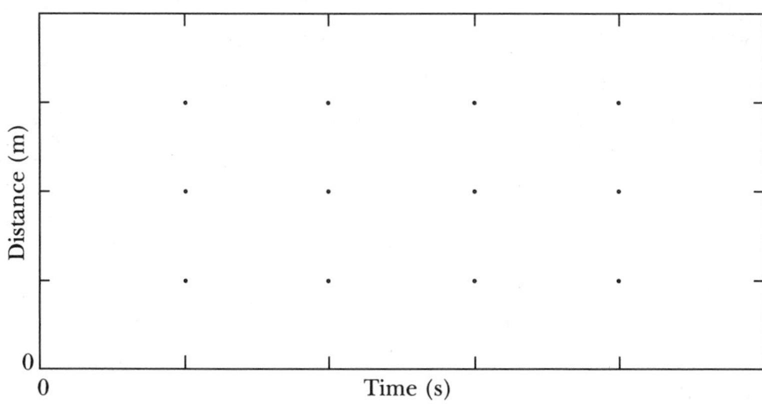

b. What is your rate of change in this case?

c. Model your motion with an expression.

When you created each of the graphs in the last task, your rate of change was constant. Consequently, the graphs representing your motion could be modeled by a line. What happens if you vary your rate of change? How does this affect the shape of the associated graph? In the next section, you will explore the answers to these questions when your rate of change varies continuously, for example, when you walk faster and faster and faster. First examine the situation where you move at one constant rate of change, followed by another constant rate, followed by another. In this case, the graph representing your motion consists of a bunch of line segments. The graph is said to be *piecewise-linear,* since each piece of the graph is itself a line. In the next task, you will convert among verbal, graphic and symbolic representations of piecewise-linear functions, and use them to model situations.

Task 1-5: Examining Piecewise-Linear Functions

1. Convert from a verbal representation to a graphical representation and model the function.

   a. Predict the graph that would be produced if you were to start at the 1-meter mark, increase your distance from the detector walking at a slow constant pace for 4 seconds, stop for 4 seconds, and then decrease your distance walking at a rapid constant pace for 2 seconds. Use a dotted line to draw your prediction on the first pair of axes given below.

   b. Compare predictions with the rest of your group. See if you can all agree. Use a solid line to draw your group's prediction on the first pair of axes. (Do not erase your original prediction.)

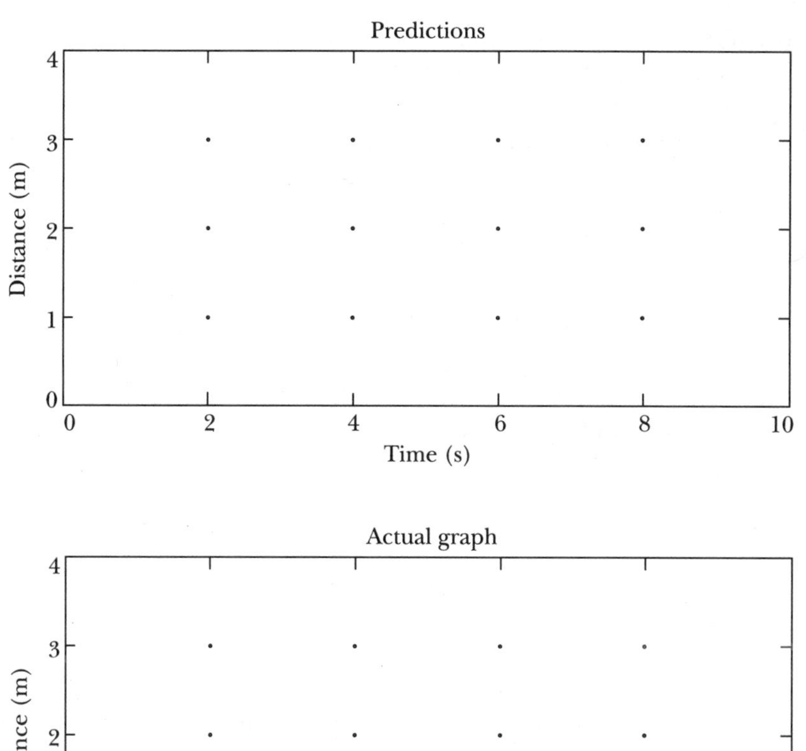

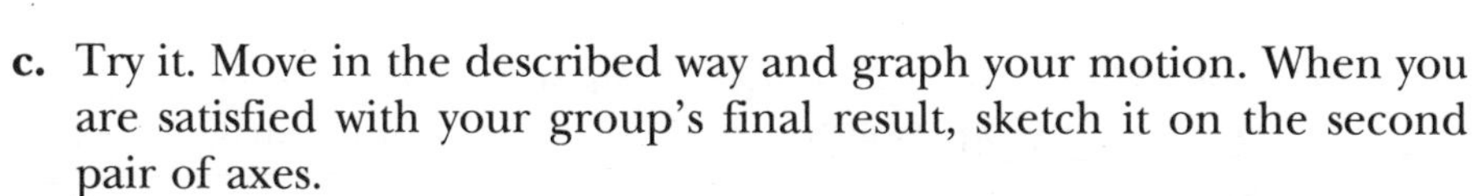

c. Try it. Move in the described way and graph your motion. When you are satisfied with your group's final result, sketch it on the second pair of axes.

d. Does your prediction have the same graph as the final result? If not, describe how you would move to make a graph that looks like your prediction.

e. For each of the ways that you moved, find an expression that represents the function during that particular time period by fitting a line to the graph. That is:

(1) Find an expression that models your movement when you started at the 1-meter mark and increased your distance from the detector by walking slowly and steadily for 4 seconds. State the domain for this piece of your function.

(2) Find an expression that models your movement when you stopped for 4 seconds. State the domain for this piece of your function.

(3) Find an expression that models your movement when you decreased your distance by walking quickly for 2 seconds. State the domain for this piece of your function.

(4) Represent your function by a *piecewise-defined* function, by combining your expressions in parts (1)–(3) using "curly bracket" notation.

2. Match a given distance versus time graph.

a. Clear any data from previous experiments. Open the file containing the **Distance Match** graph. The distance versus time graph shown below will appear on your screen.

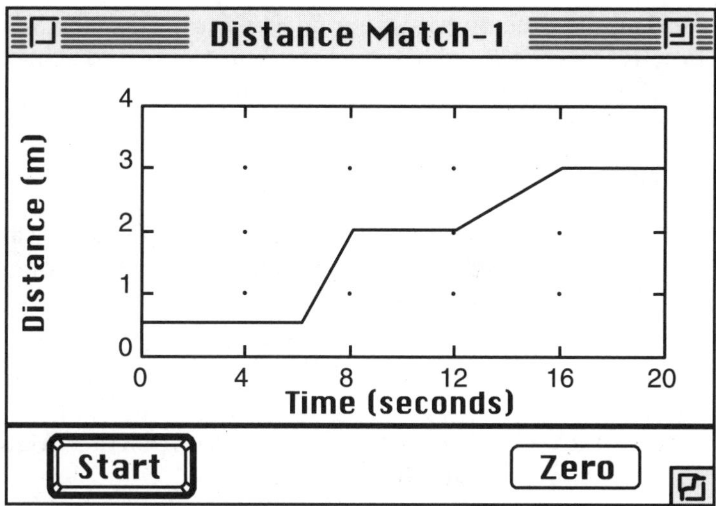

b. Try to duplicate the **Distance Match** graph. Sketch your best match on the axes given above.

*Note: The **Distance Match** graph is stored in the computer as **Data B**. New data from the motion detector are always stored in **Data A** and therefore can be collected without erasing the given graph.*

c. Complete the following verbal description of the motion depicted by the **Distance Match** function.

First you stand still at a distance of _____ meters for _____ seconds. Then you ...

d. During which time interval is the rate of change the greatest?

e. Analyze the shape of the **Distance Match** function. Express your conclusions using open interval notation.

(1) Find all the time intervals (if any) where the **Distance Match** function is *increasing*—that is, find the intervals where you increased your distance from the detector as you tried to duplicate the graph.

(2) Find all the time intervals (if any) where the **Distance Match** function is *decreasing*—that is, where you decreased your distance from the detector.

(3) Find all the time intervals (if any) where the **Distance Match** function is *constant*—that is, where your distance from the detector was constant.

f. Use "curly bracket" notation to represent the **Distance Match** function by a piecewise-defined function.

3. Explain why a person cannot replicate each of the following graphs by walking in front of the motion detector.

a.

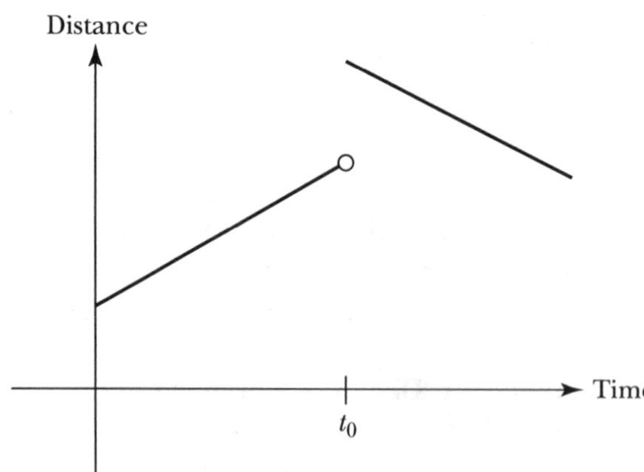

b.

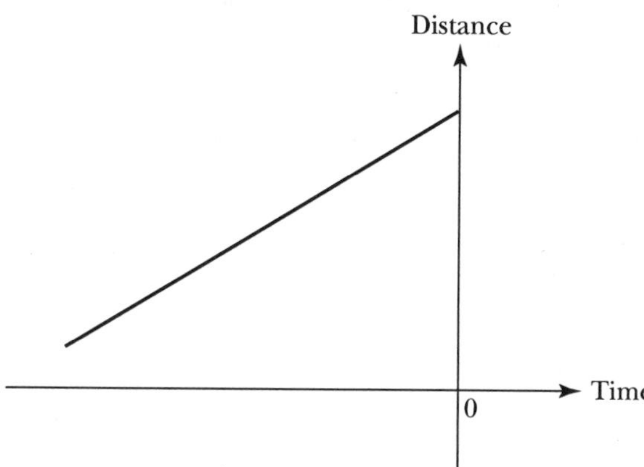

c.

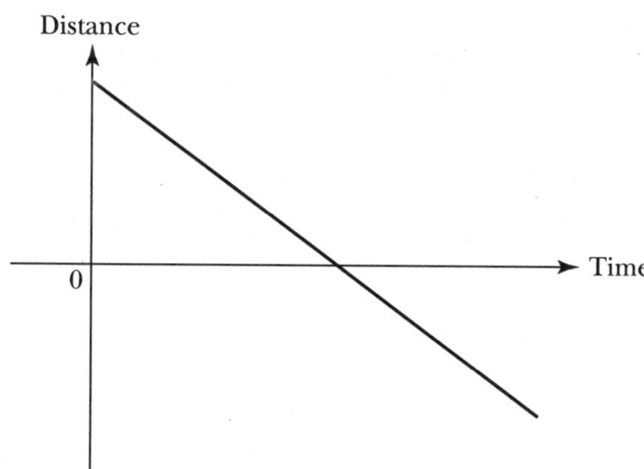

4. For each of the following scenarios, give the general shape of a distance versus time graph that represents the situation. Label the time of specific events, such as "stopped to tie shoe," "stepped in puddle," "ran to class," on the horizontal axis. Label specific locations, such as "leave dorm," "arrive at class," along the vertical axis.

a. You left your dorm and started walking slowly to class when you realized that you had forgotten your books, so you ran back to your dorm, grabbed your books and sprinted to class.

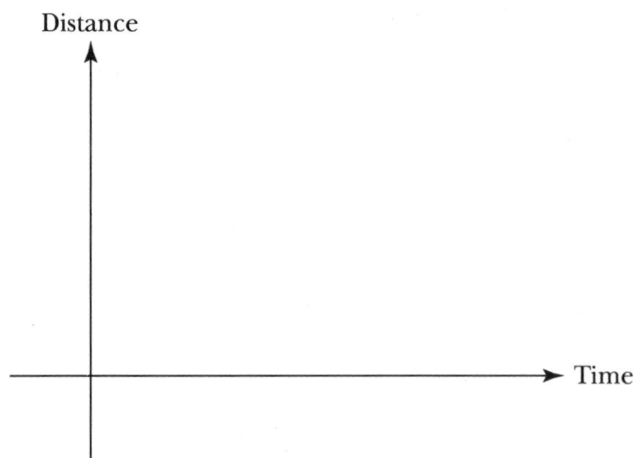

b. As you were walking slowly to lunch, you realized that you were late for your meeting with Christa so you started to run.

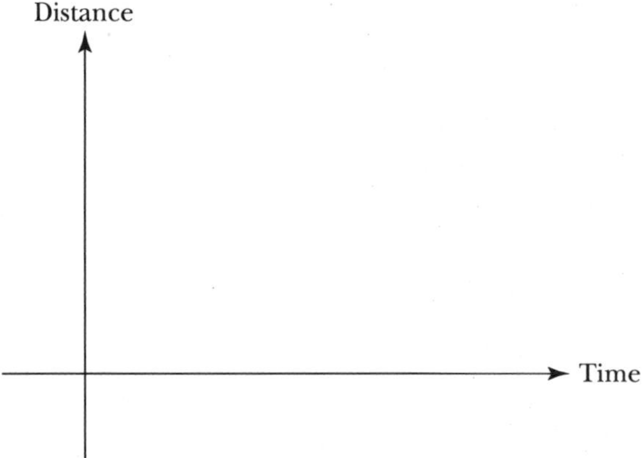

c. On your way to soccer practice you ran into Greta. You stopped to chat before continuing on to the gym.

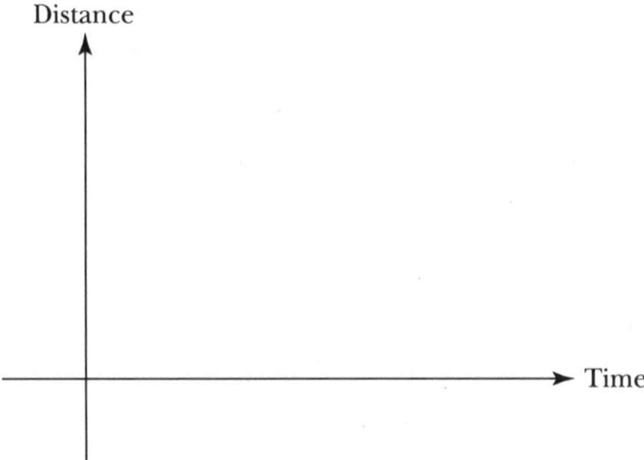

5. For each of the following distance versus time graphs, describe a situation that the graph might represent.

a.

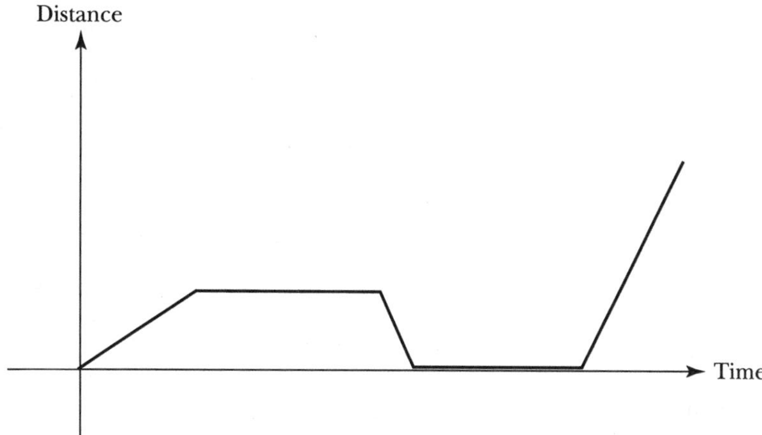

b.

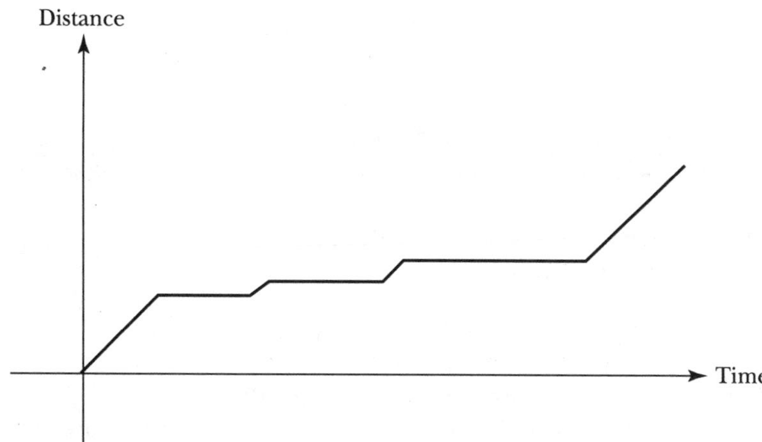

c.

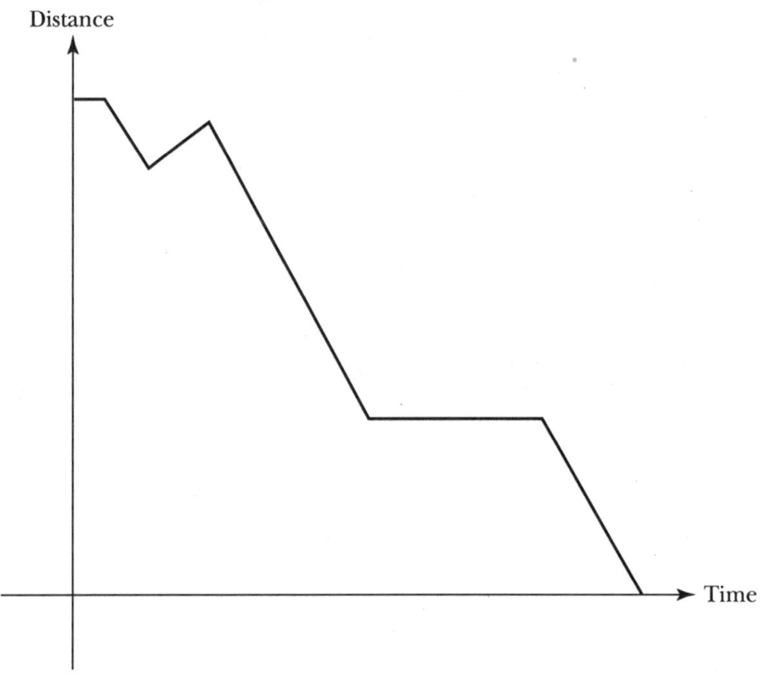

Unit 1 Homework After Section 2

- Complete the tasks in Section 2 in the Activity Guide. As usual, be prepared to discuss them in class.
- Analyze the shape of some functions in HW1.2.

HW1.2 For each of the following graphs, identify where the graph is increasing, decreasing, and constant. Note: Label the appropriate points on the horizontal axis *a*, *b*, *c*, and so on. Use these labels to express your answers using open interval notation.

1.

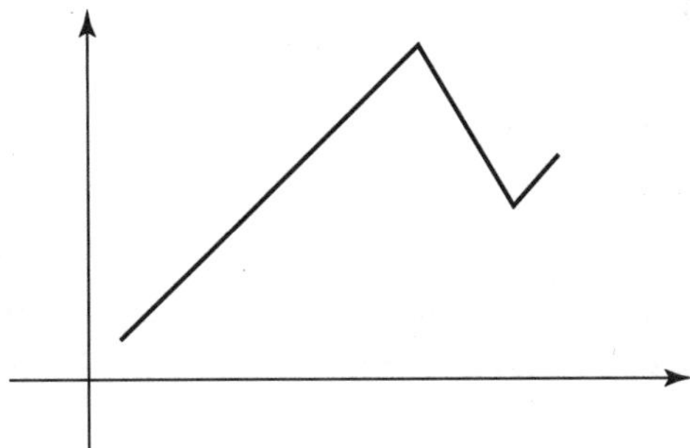

2.

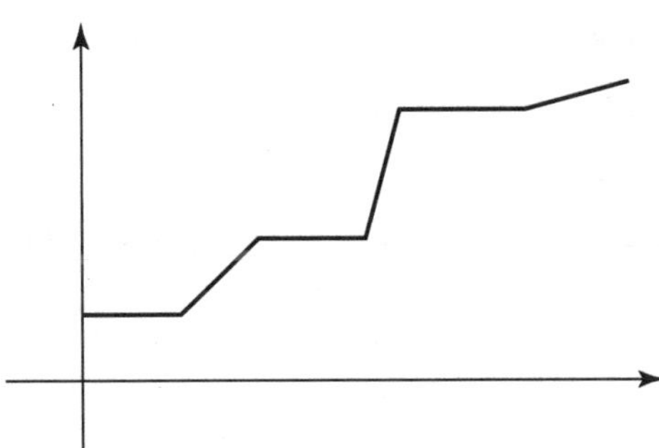

3.

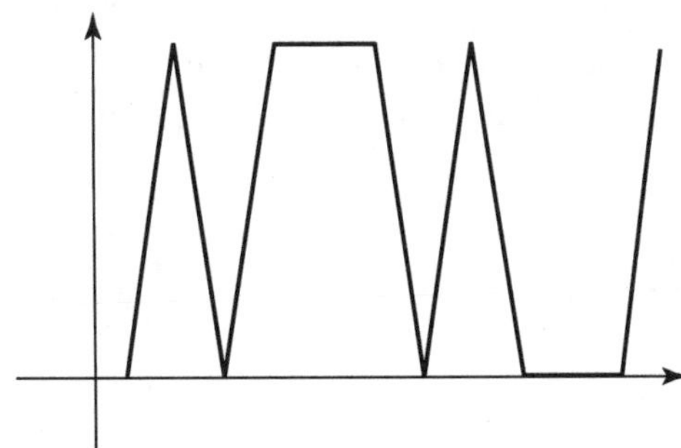

• Model some situations where you know the exact rate of change in HW1.3.

HW1.3 Consider the following situations.

1. You start at the 1-meter mark. For 3 seconds you increase your distance from the detector at the rate of 2 meters per second. You stand still for 4 seconds.

a. Draw a precise distance versus time graph corresponding to this situation. Carefully label the axes.

b. Model your motion by a piecewise-defined function.

c. Fill in the following table.

Time	0	1	2	2.5	3	5.5	7
Distance							

2. Starting at the 10-meter mark you walk at a constant rate of -2 meters per second. In other words, you decrease your distance by 2 meters per second.

a. What is your distance when $t = 3$ seconds?

b. At what time are you 1 meter from the detector?

c. How many seconds must you walk before you reach the detector?

d. Model your motion with a linear expression.

e. In this example the rate of change is negative. What does this tell you about the behavior of the graph of the associated distance versus time graph?

- Explain what really happens in HW1.4.

HW1.4 Actually, we have been a little sloppy with our graphs. It is not possible to duplicate exactly a graph, such as the one given below, by walking in front of the motion detector. There are two places where we have difficulty: the initial portion and the corner (peak), both of which are circled. Explain what actually happens at these places when you try to duplicate them by walking in front of the motion detector. Make a more realistic sketch.

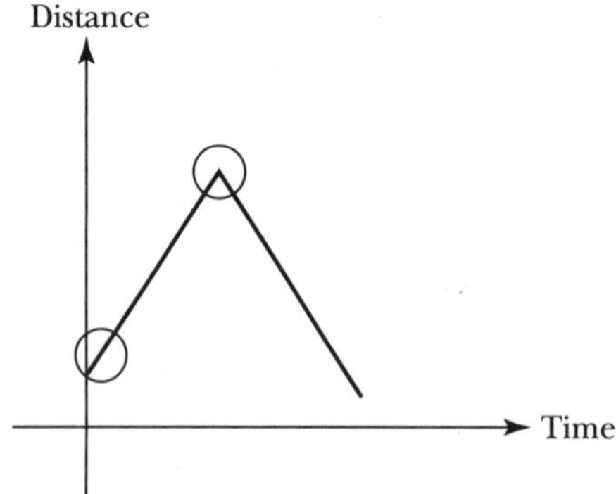

- Examine the concept of independent and dependent variables in HW1.5.

HW1.5 When creating a distance versus time graph, your distance is determined by the value of the time—that is, your distance "depends on" the time or your distance is a "function of" time. Consequently, distance is called the *dependent* variable, whereas time is called the *independent* variable. In general, the quantity which varies freely is the independent variable, whereas the quantity that depends on the independent variable is the dependent variable.

For each of the following functional descriptions:

i. Identify the two quantities that vary. Give each quantity a variable name and indicate the unit of measurement for the variable. For example, time (seconds) or speed (miles/hour).

ii. Determine which quantity is the independent variable and which is the dependent variable.

iii. Illustrate the basic relationship between the variables by sketching the general shape of the graph. Label each axis with its variable name. Note: It is customary to place the independent variable on the horizontal axis and the dependent variable on the vertical axis.

1. The temperature rises from 6 A.M. until 6 P.M. and then falls from 6 P.M. to 6 A.M.
2. The number of bacteria in the petri dish doubles every 2 hours.
3. Dan works at his office 12 hours on Monday and Friday, 4 hours on Tuesday and Thursday, and 8 hours on Wednesday. He does not work at his office over the weekend.
4. The water continually rises until high tide occurs and then falls until low tide occurs.
5. The higher the cost of an item, the fewer people are willing to buy.
6. As you increase the rate at which you do push-ups, your heart rate increases.

- In your past study of mathematics, you may have used the "Vertical Line Test" to determine if a given graph represents a function. Explain why this test makes sense in HW1.6. Keep in mind that for each input, a function returns exactly one output. No more. No less.

HW1.6 The Vertical Line Test claims:

> A graph represents a function if and only if every vertical line in the xy-plane intersects the graph in at most one point.

Explain why this makes sense. State your explanation in terms of "inputs" and "outputs," and support it with two diagrams, one which represents a function and one which does not.

• Use the Vertical Line Test in HW1.7.

HW1.7 For each of the following graphs, use the Vertical Line Test to determine if the graph represents a function. If the graph fails the test, draw a vertical line which shows that the graph does not have one output for each input.

> *Notes: An open circle indicates a "hole" in the graph, whereas a closed circle indicates a point on the graph. HW1.7 may be written on, even though the general directions for homework ask you not to scrunch your answers onto the homework pages.*

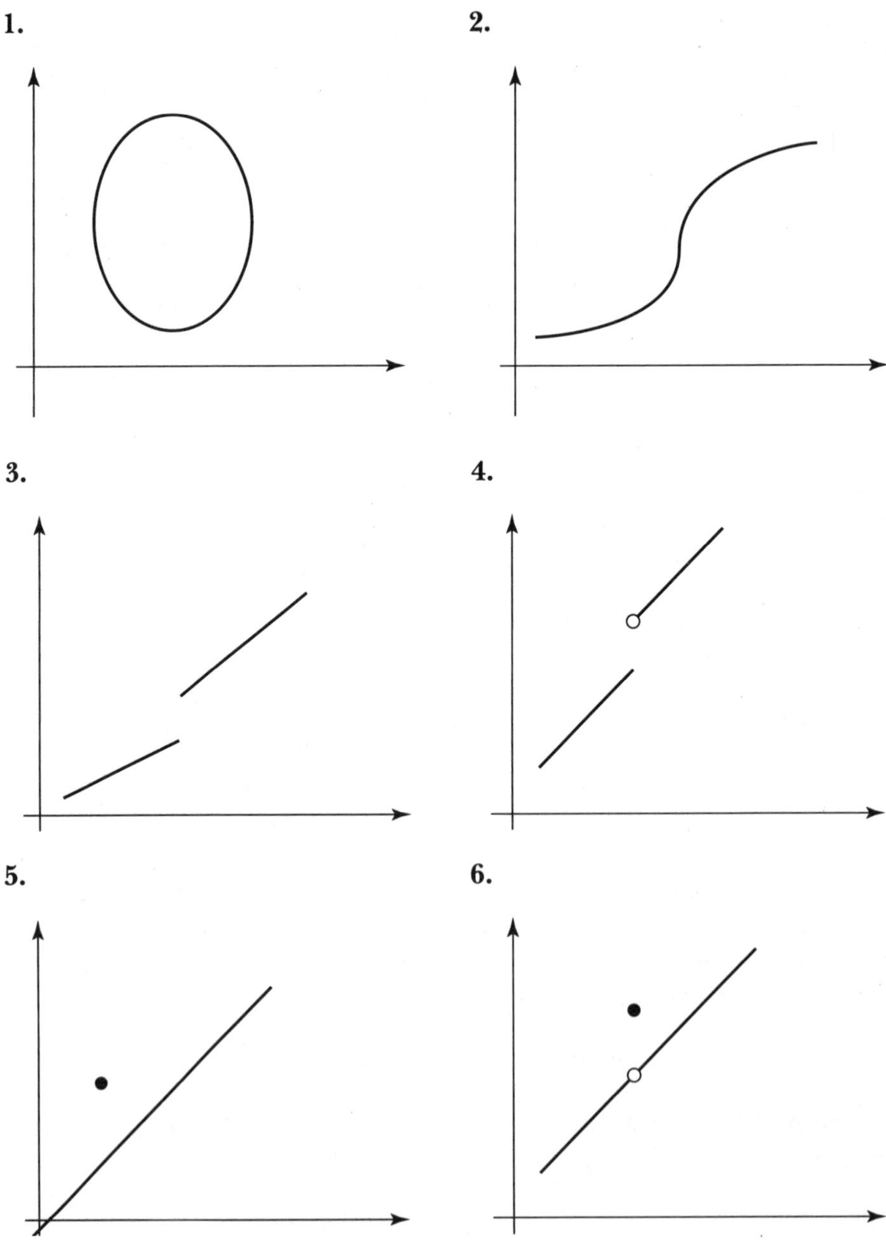

7.

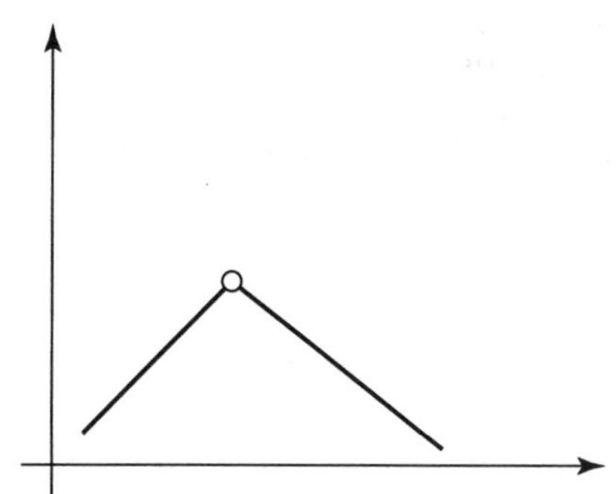

8.

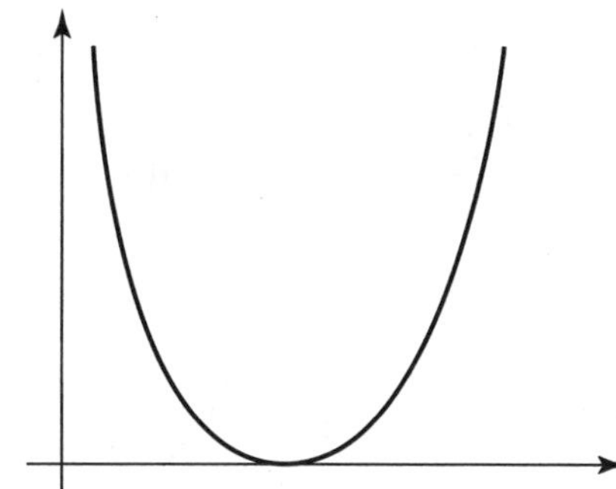

9.

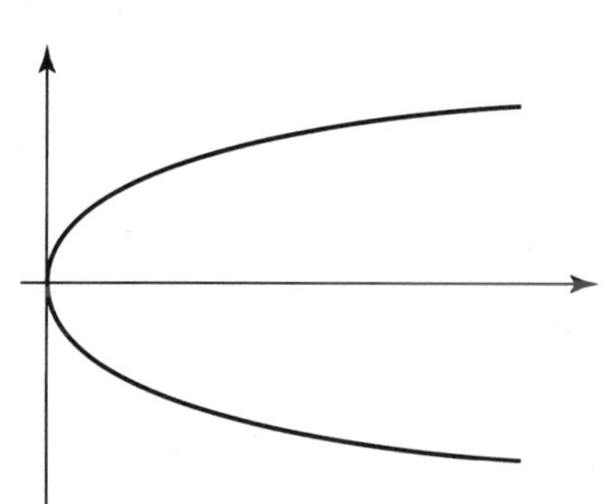

10.

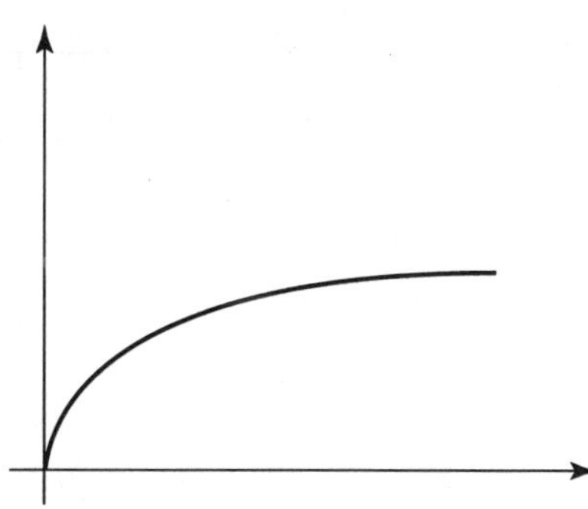

11.

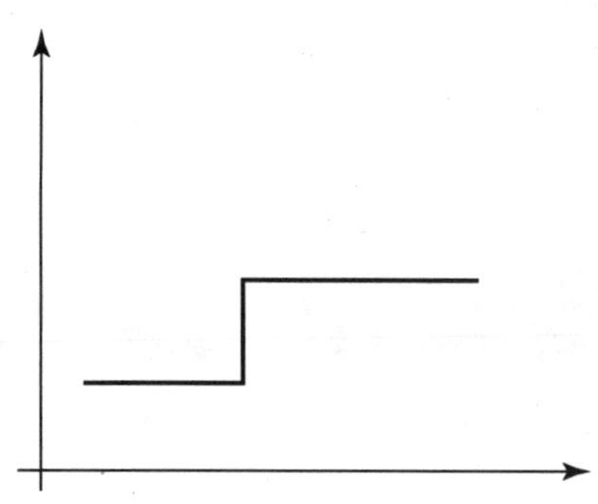

12.

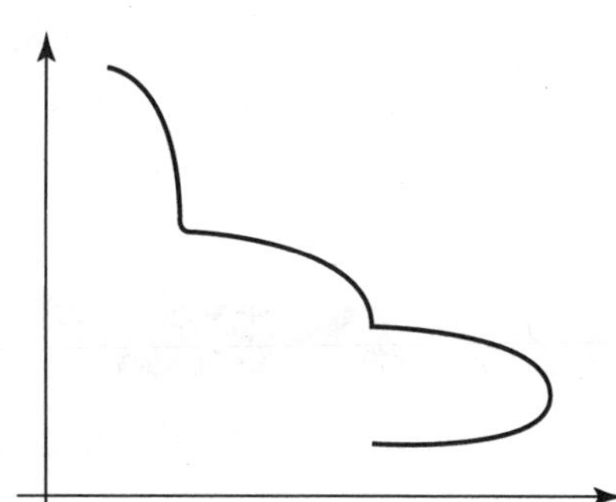

• Examine some piecewise-defined linear functions in HW1.8.

HW1.8 Examine some piecewise-linear functions that have "jumps."

1. Represent the following function using curly bracket notation. Be careful at the end-points of the subdomains.

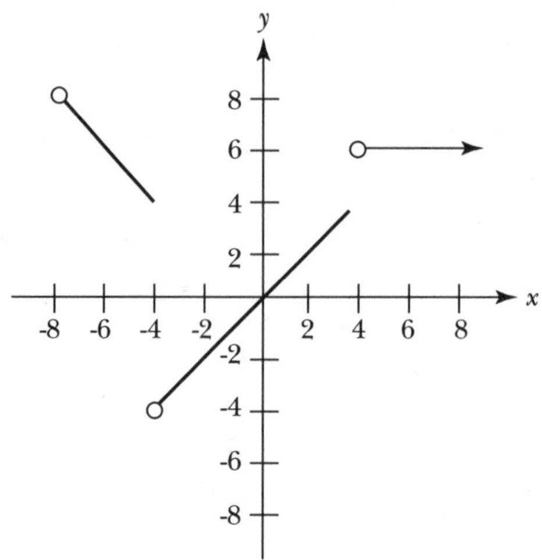

2. Sketch the following function. As usual, use an open circle to denote an end-point which is not on the graph.

$$s(t) = \begin{cases} -t + 3, & \text{if } -1 \leq t \leq 1 \\ t + 3, & \text{if } 1 < t < 4 \\ t, & \text{if } 4 \leq t < 10 \end{cases}$$

SECTION 3

Analyzing Smooth Curves

In the last section, you created linear functions by walking at a constant pace—that is, by keeping your rate of change constant. You observed that your rate of change, or velocity, corresponded to the slope of the line which modeled your movement. When you increased your distance from the detector, your rate of change was positive and you created a line with a positive slope. Conversely, when you decreased your distance, your rate of change was negative and you created a line with a negative slope.

In this section, you will look at functions whose graphs bend and wiggle, but which do not have any sharp corners or jumps. These graphs are called *smooth curves.* As you probably guessed, you can create a smooth curve by varying your rate of change—for instance, by walking faster and faster or by walking back and forth in front of the detector. The questions in which you are interested are: How does varying your rate of change affect the shape of the curve? How do you move to create a curve that opens up, opens down, or turns? How can you measure your rate of change at a given instant?

One way to answer these questions is to analyze the behavior of the *tangent line* to a curve. In the next task, you will develop an intuitive understanding of what a tangent line to a curve is. In subsequent tasks, you will use the motion detector to explore the relationship between the tangent line and a smooth curve.

Task 1-6: Developing an Intuitive Understanding of a Tangent Line to a Curve

1. Begin by thinking about a familiar situation: a tangent line to a circle. A *tangent line to a point on a circle* is the line that touches the circle at exactly that one point, even though the line extends indefinitely in both directions.

   a. Carefully sketch the tangent line to each of the circles given below at the point P.

   (1)

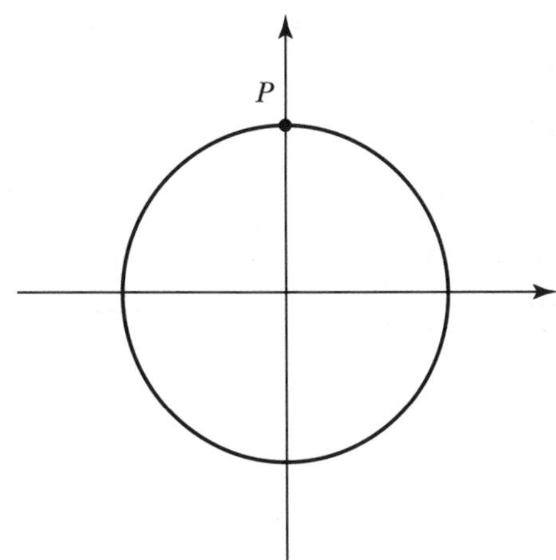

(2)

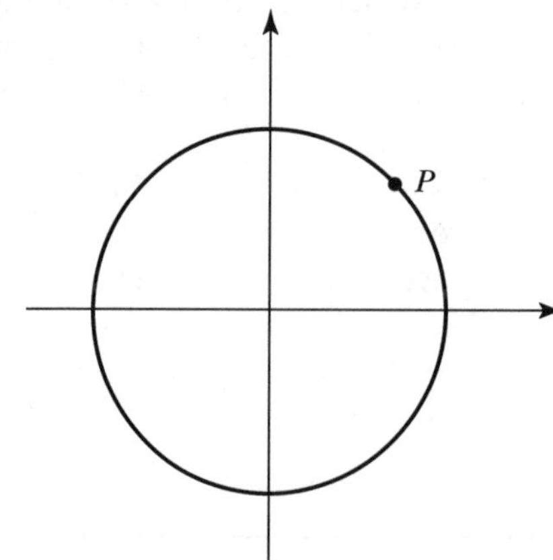

(3)

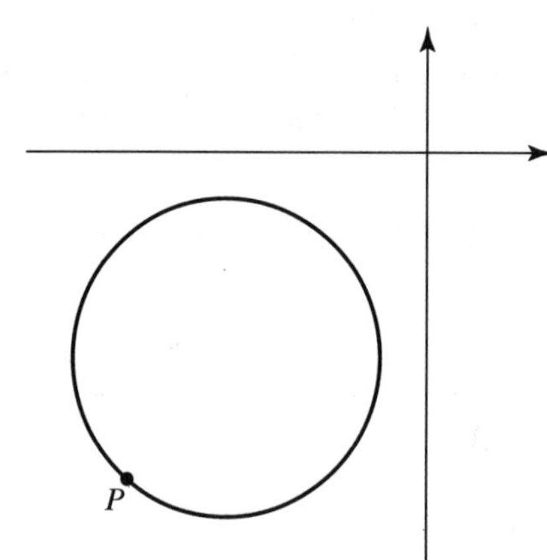

b. In each of the following diagrams, explain why the line L is *not* a tangent line to the given circle.

(1)

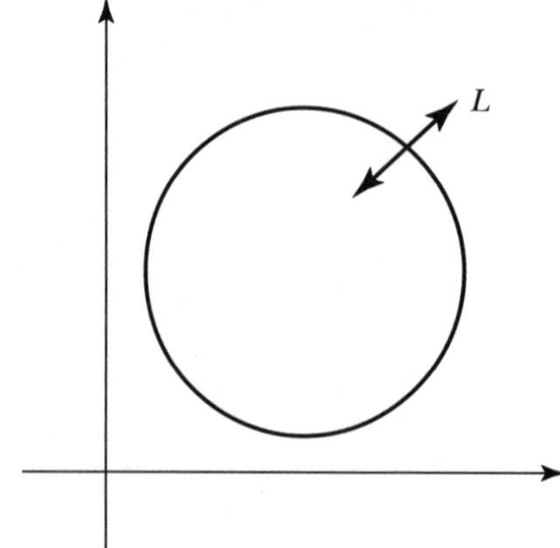

(2)

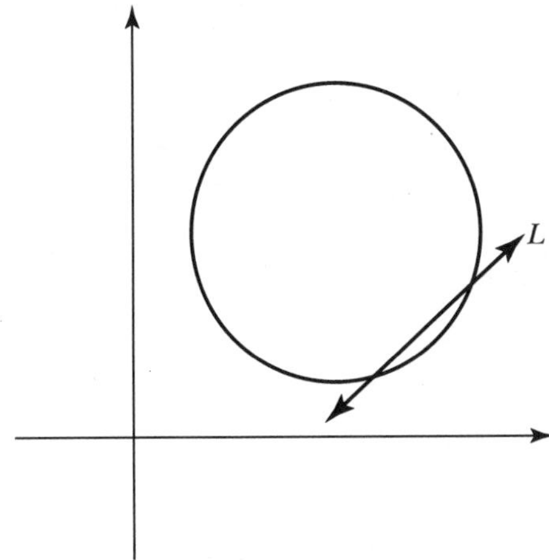

c. Describe a way of constructing the tangent line to a circle at a given point.

d. Another way to identify the tangent line at a point on a circle is to repeatedly "zoom in" and magnify a small region containing the point. As you get closer and closer, the tangent line and the part of the circle near the point merge.

Convince yourself that this is the case by repeating the following zoom-in-and-magnify process: Draw a small box around *P*, cover the portion of the graph outside of the small box, and then copy the contents of the box into a larger box nearby. The diagram given below illustrates the process for the first time. Repeat this zoom-in-and-magnify process two more times.

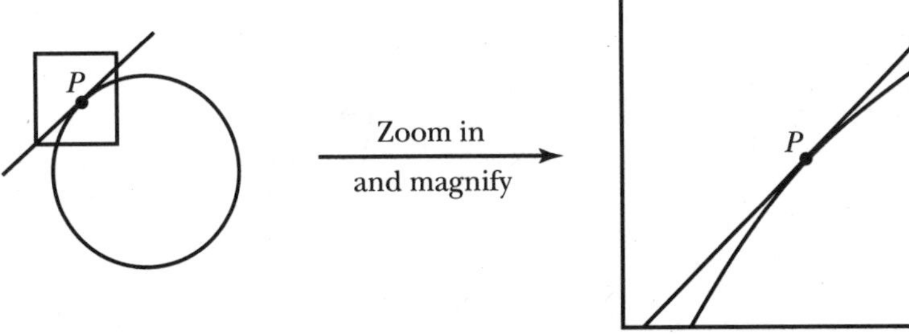

2. In part (1) you examined the idea of a tangent line to a circle. Next extend this notion to a smooth curve whose graph bends and wiggles.

One way to find the tangent line at a given point P on a smooth curve is to repeatedly zoom in and magnify the portion of the curve containing P. As you get closer and closer, the graph of the smooth curve will "straighten out" and merge with the graph of the tangent line to the curve at P. This, of course, is exactly what happens in the case of the tangent line to a circle.

a. Identify the tangent line to the graph of the function at the point P in the diagram below. Use a zoom-in-and-magnify argument to justify your choice.

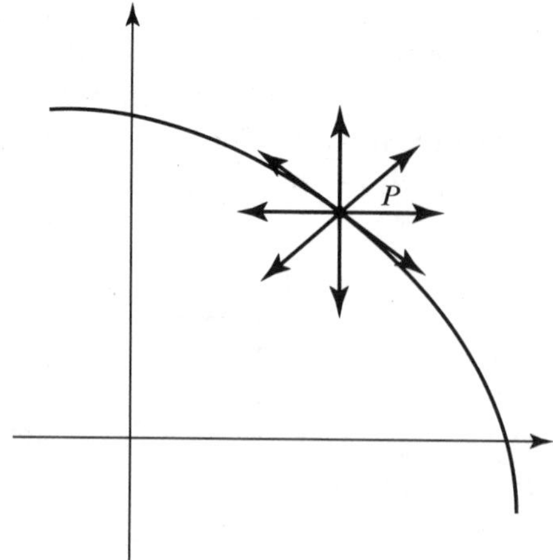

b. Carefully sketch the tangent line to each of the smooth curves given below at the point P. In each case indicate the *sign* of the slope of the tangent line—that is, indicate whether the value of the slope is a positive number (+), a negative number (−), zero (0), or undefined.

(1)

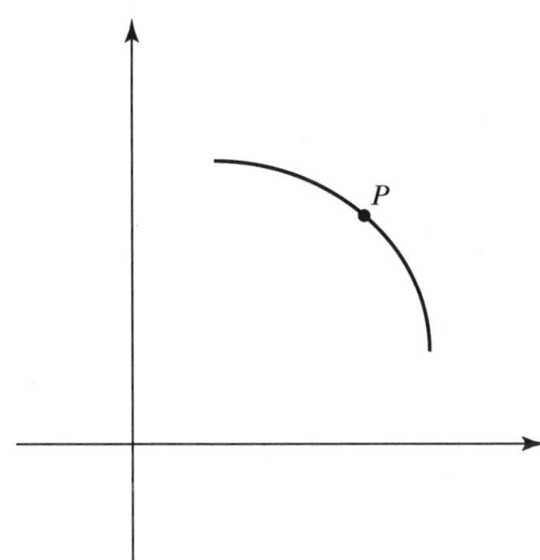

Sign of the slope of the tangent line at P:

(2)

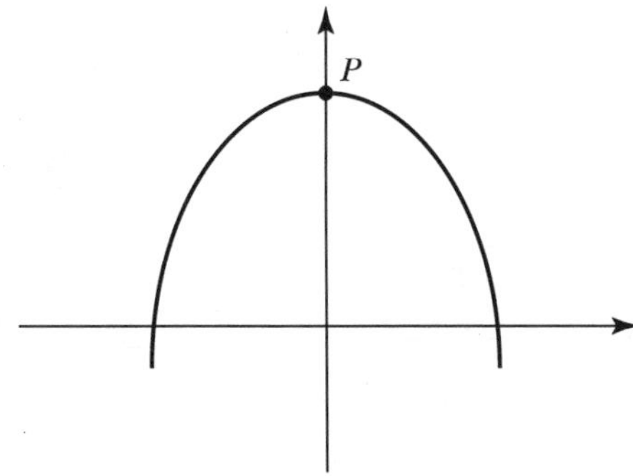

Sign of the slope of the tangent line at P:

(3)

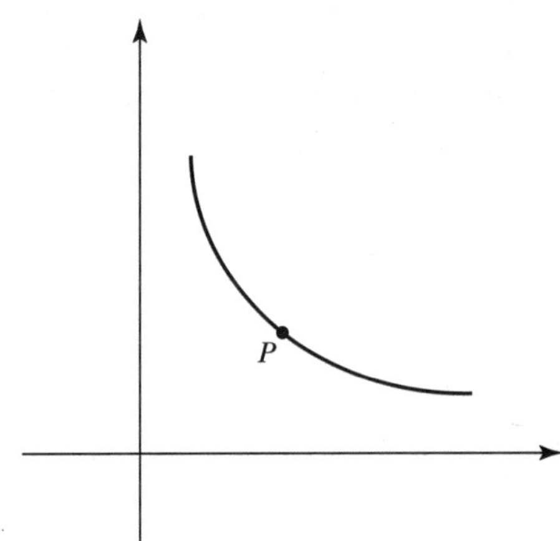

Sign of the slope of the tangent line at P:

(4)

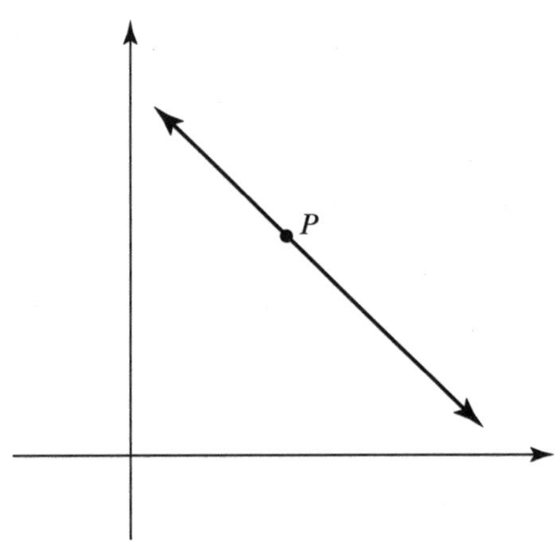

Sign of the slope of the tangent line at P:

(5)

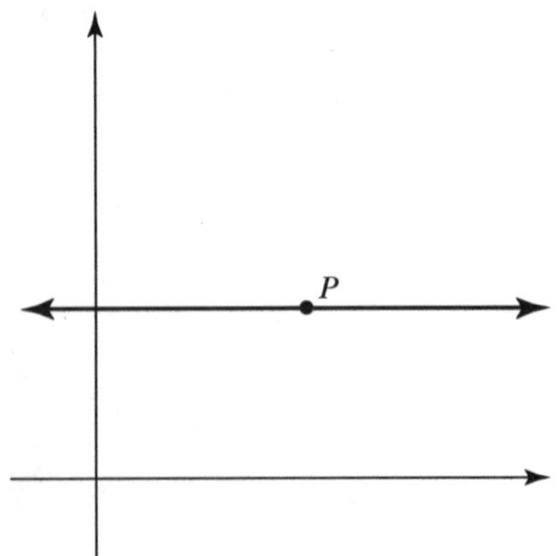

Sign of the slope of the tangent line at P:

(6)

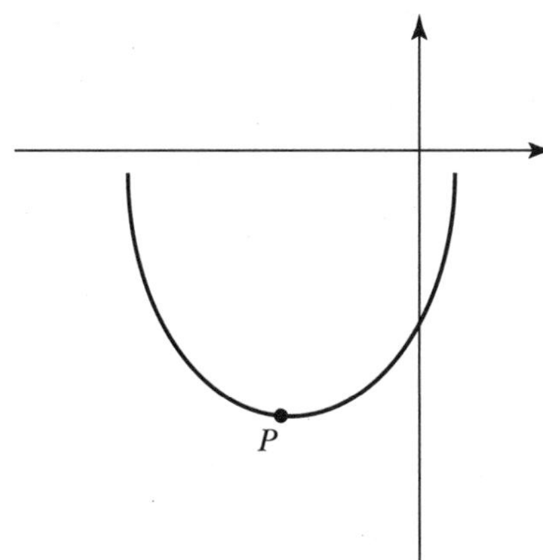

Sign of the slope of the tangent line at P:

(7)

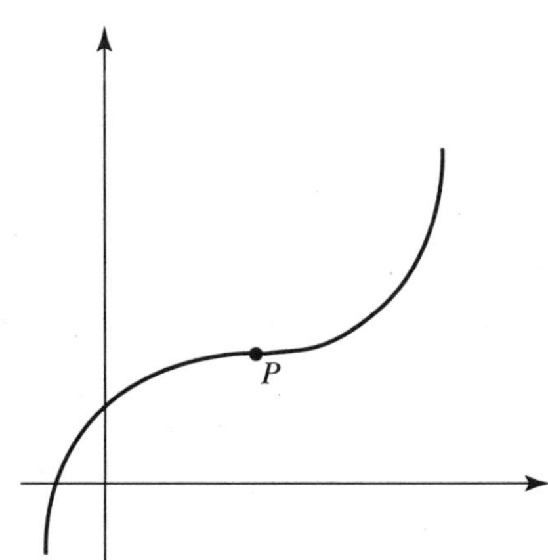

Sign of the slope of the tangent line at P:

(8)

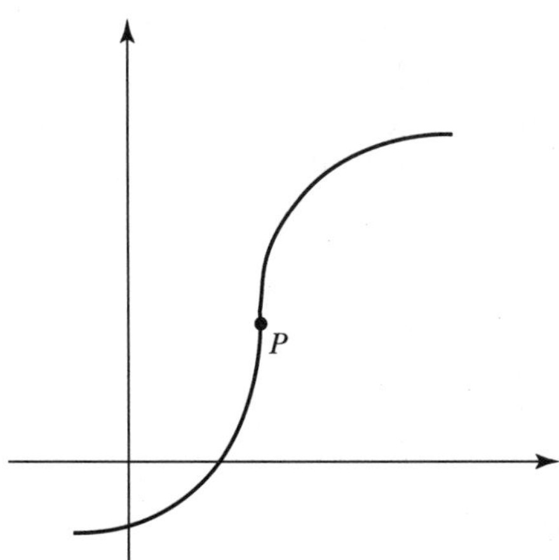

Sign of the slope of the tangent line at P:

c. The tangent line at a point P on a circle intersects the circle at exactly one point, namely P. The tangent line to a smooth curve, however, may intersect the curve at more than one point. Illustrate this fact with a sketch; that is, draw a smooth curve that passes through P, where the tangent line at P intersects the curve at at least one point other than P:

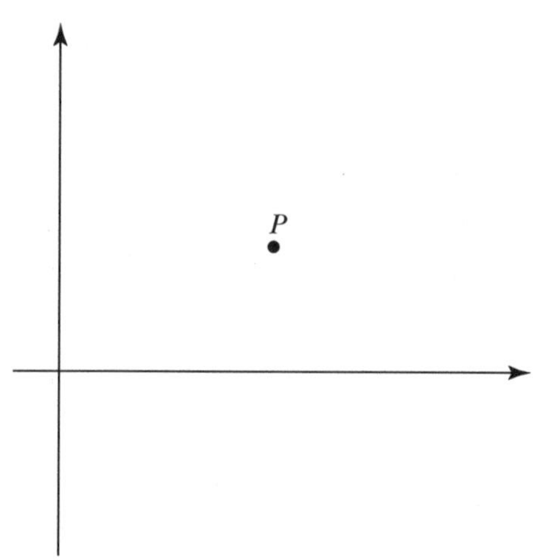

d. If a tangent line exists at a point then it is unique—that is, there is one and only one. Use a zoom-in-and-magnify argument to explain why this is a reasonable conclusion.

e. Tangent lines do not exist at sharp peaks and dips.

Zoom in and magnify a tip of the dude's hair. What happens in this case? Based on your observations and on the properties of tangent lines, explain why it is reasonable to conclude that tangent lines do not exist at sharp peaks and dips.

The slope of the tangent line to a smooth curve gives the *instantaneous rate of change* of the function at the point *P*. This is a very important value, since it indicates the rate at which the function is changing. For example, suppose you planted two different types of plants, Plant #1 and Plant #2, and graphed their heights over a given time period using the same scales for each pair of axes.

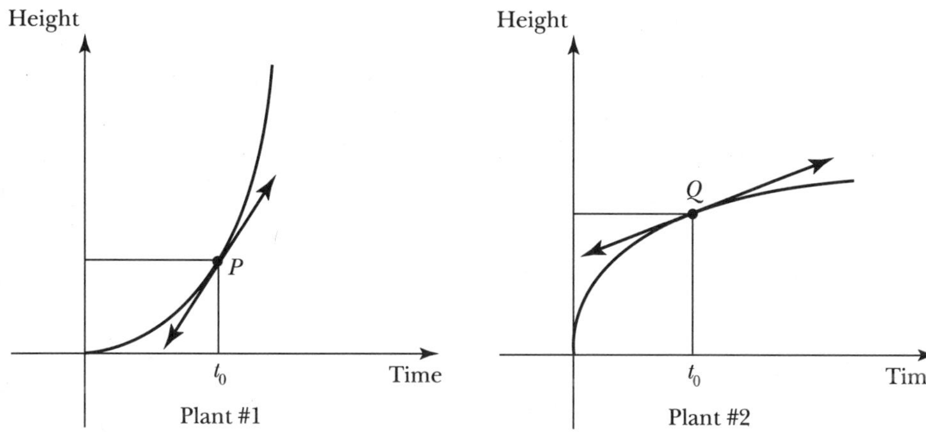

The slope of the tangent line to each point on the curve indicates how fast the plant is growing at that time. Look closely at what happens at time t_0. Plant #1 is growing more rapidly than Plant #2, since the slope of its tangent line, and hence its rate of change, is greater. On the other hand, it is interesting to note that even though Plant #1 is growing faster, it is still shorter than Plant #2 at time t_0. (You will have a chance to explain what is happening here in the homework exercises.)

Being able to calculate the rate of change of a function at a given point is very useful. There's one big problem, however. You do not know how to find the actual value of the slope of the tangent line to a smooth curve! In order to use the slope formula which you learned in algebra, you need two points, and in the case of a tangent line to a curve, you have only one, namely *P*. The good news is that calculus will come to your rescue and provide you with a way of doing this. In the meantime, you can use your intuitive understanding of tangent line which you developed in the last task. You know the circumstances under which it exists. You know what it looks like, and given a point, you can sketch a reasonable approximation. The motion detector can help you develop a feel for what rate of change means, since its value depends on how fast you are moving and its sign depends on whether you are increasing or decreasing your distance from the detector. In the following task, you will use the motion detector to discover some of the relationships between the shape of a graph and the behavior of the tangent line as it moves along the curve.

Task 1-7: Investigating the Behavior of the Tangent Line near a Turning Point

1. Create a smooth curve by walking back and forth in front of the motion detector, changing your direction at least two times. Print a copy of your graph and place it in the space below.

2. Enter the **Analyze Data** mode.

   a. Consider the time intervals where your function is increasing.

   (1) Below the horizontal axis on your graph, label each interval where your function is increasing.

   (2) Express each interval using open interval notation.

   b. Consider the time intervals where your function is decreasing.

   (1) Below the horizontal axis on your graph, label each interval where your function is decreasing.

   (2) Express each interval using open interval notation.

   c. Find the first and second coordinate of each point where you changed direction—that is, find the coordinates of the *turning points* of your graph.

d. Consider the turning points where your graph opens down—that is, where your function has a *local maximum.*

(1) Mark each point on your graph where your function has a local maximum.

(2) Describe the behavior of your function to the left and right of the local maximum in terms of increasing and decreasing.

e. Consider a turning point where your graph opens up—that is, where your function has a *local minimum.*

(1) Mark each point on your graph where your function has a local minimum.

(2) Describe the behavior of your function to the left and right of the local minimum in terms of increasing and decreasing.

3. Turn on the **Tangent** line. Analyze the behavior of the tangent line to your graph by moving the tangent line along your graph.

a. What can you conclude about the sign of the slope of the tangent line in the intervals where your function is increasing?

b. What can you conclude about the sign of the slope of the tangent line in the intervals where your function is decreasing?

c. What can you conclude about the value of the slope of the tangent line at each of the turning points on your graph?

d. Consider a turning point where your function has a local maximum. Zoom in on a region containing the point.

(1) Sketch the general shape of a smooth function near a local maximum. Sketch a tangent line to the left, to the right, and at the turning point.

(2) Describe the behavior of the slope of the tangent line as it moves along the curve from left to right, through the local maximum. In other words, what happens to the rate of change near a local maximum?

e. Reset the axes. Consider a turning point where your function has a local minimum. Zoom in on a region containing the point.

(1) Sketch the general shape of a smooth function near a local minimum. Sketch a tangent line to the left, to the right, and at the turning point.

(2) Describe the behavior of the slope of the tangent line as it moves along the curve from left to right, through a local minimum. In other words, what happens to the rate of change near a local minimum?

When a smooth function is increasing, its rate of change is positive, the slope of the tangent line to the curve is positive, the graph rises from left to right, and the function is growing. On the other hand, when a smooth function is decreasing, its rate of change is negative, the slope of the tangent line to the curve is negative, the graph falls from left to right, and the function is decaying. At a local extremum—that is, at a turning point—the rate of change is zero, thus the slope of the tangent line is zero. This provides a lot of information about the rise and fall of the curve, but does not tell you anything about the *concavity* of a function—that is, when it is shaped like

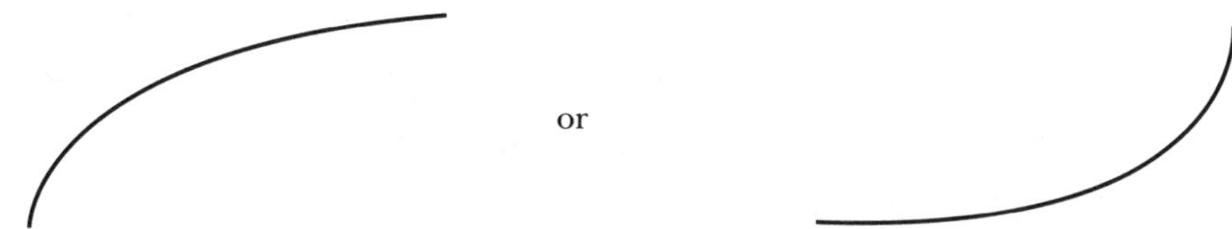

Notice that both these functions are increasing. The curve on the left, however, opens down and is said to be *concave down,* whereas the one on the right opens up and is said to be *concave up.*

In the next task you will examine the relationship between the concavity of a function and your velocity when you move in front of a motion detector. You will examine also how the concavity of a function is related to the location and behavior of the tangent line as it travels along its graph.

Task 1-8: Contemplating Concavity

1. Examine the location of the curve with respect to the tangent line. Does the curve lie above the tangent line or below?

   Consider the following graphs:

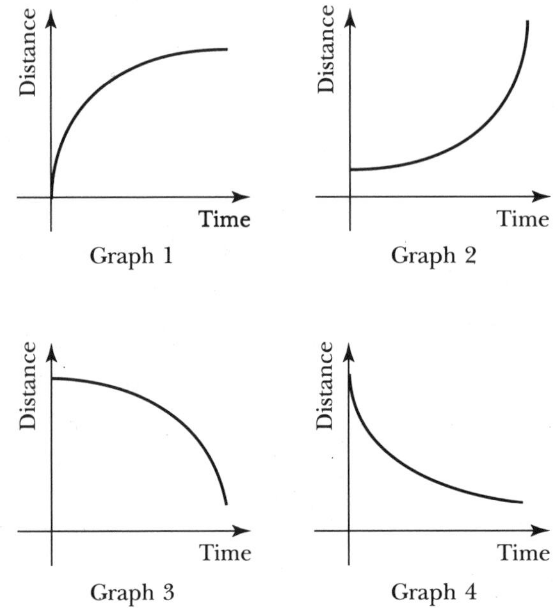

   a. Pick a point on each graph and sketch the tangent line at that point.

   b. Graphs 2 and 4 are concave up. Describe the location of the curve with respect to the tangent line in both of these cases.

   c. Graphs 1 and 3 are concave down. Describe the location of the curve with respect to the tangent line in both of these cases.

2. Describe how you must move to produce each of the distance versus time graphs shown above.

   a. Graph 1:

   b. Graph 2:

c. Graph 3:

d. Graph 4:

3. Use the motion detector to check your responses to part 2. Create distance versus time graphs that have the same shape as Graphs 1-4. If necessary, modify your descriptions in part 2.

4. One way to think about a tangent line to a graph is think about a teeny-tiny bug crawling along the curve with a pencil strapped to its back. The pencil corresponds to the tangent line at each point where the bug touches the curve.

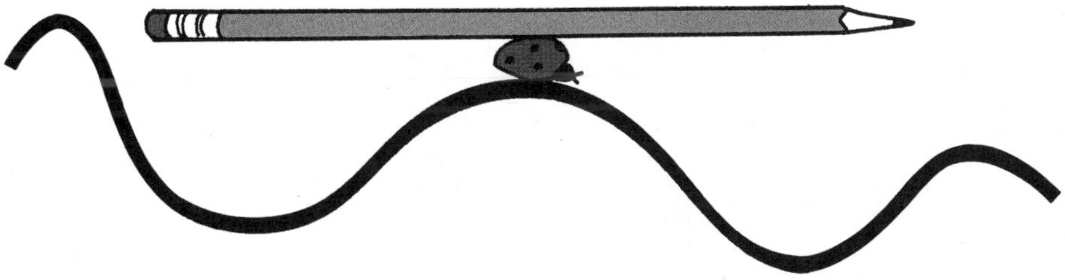

a. Label the intervals on the graph given above where the function is concave up and where it is concave down.

b. Describe the location of the ant (and the pencil) as the bug crawls along a portion of the curve that is concave up.

c. What happens to the bug (and the pencil) when the curve changes concavity, from up to down or down to up?

Note: This is called an inflection point.

d. Mark all the local maxima on the curve. Describe the concavity at a local maximum.

e. Mark all the local minima on the curve. Describe the concavity at a local minimum.

5. Sketch a graph of some basic situations. For each of the following:

(i) Sketch a graph satisfying the specified conditions.

(ii) Draw some typical tangent lines.

a. The function is increasing. The graph is concave up.

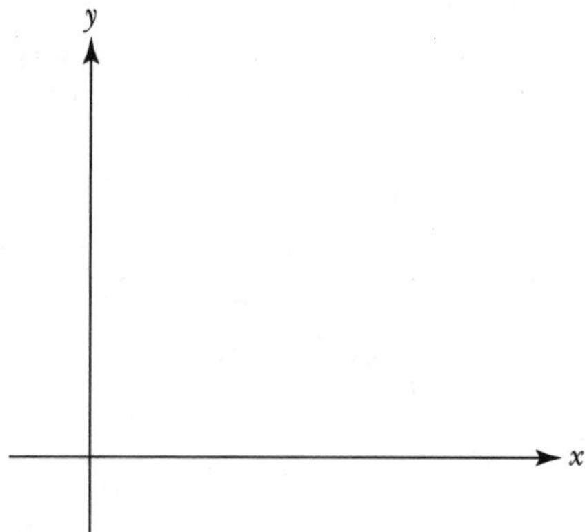

b. The function is decreasing. The graph is concave up.

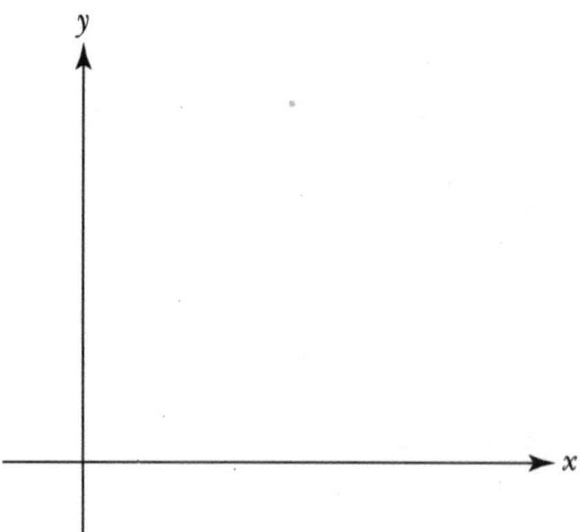

c. The function is increasing. The graph is concave down.

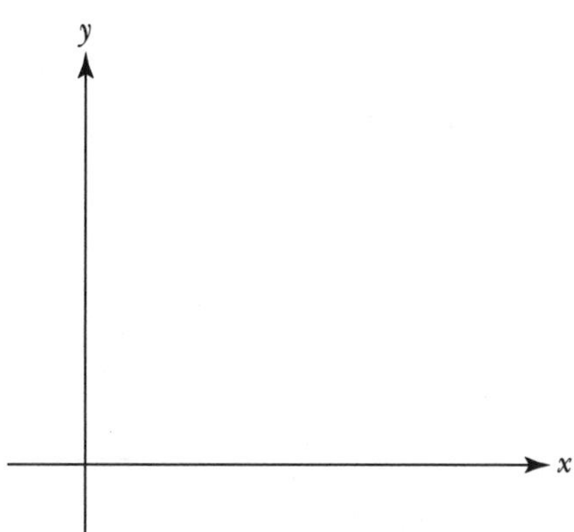

d. The function is decreasing. The graph is concave down.

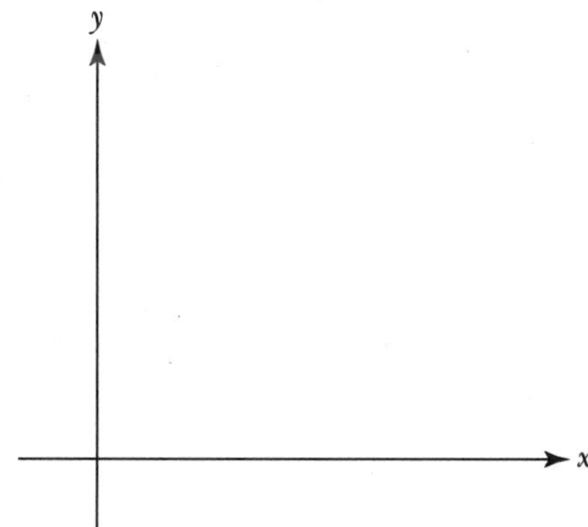

e. The function is increasing. The tangent line at t_0 cuts through the graph of the function.

Note: The point where the tangent line passes through the graph is an inflection point.

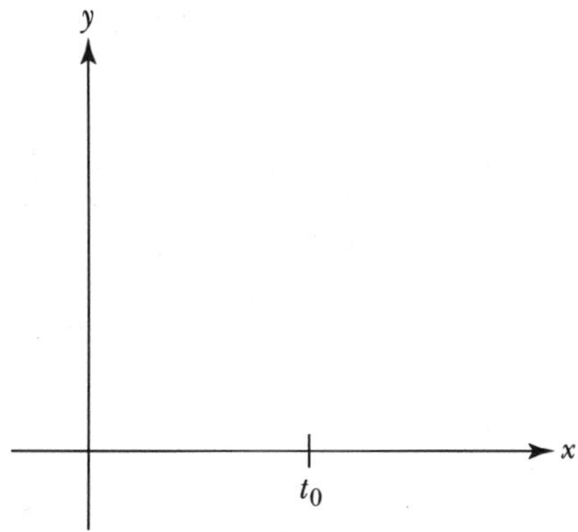

f. The function is decreasing. t_0 is an inflection point.

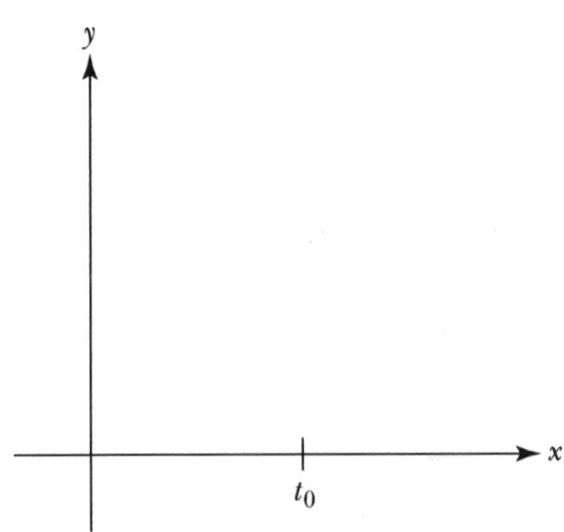

g. The tangent lines at points to the left of t_0 have negative slopes. The slope of the tangent line at t_0 is 0. The tangent lines at points to the right of t_0 have positive slopes.

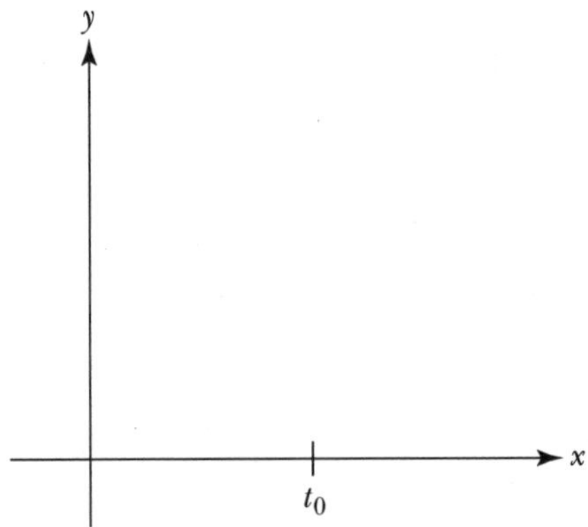

h. The tangent lines at points to the left of t_0 have positive slopes. The slope of the tangent line at t_0 is 0. The tangent lines at points to the right of t_0 have negative slopes.

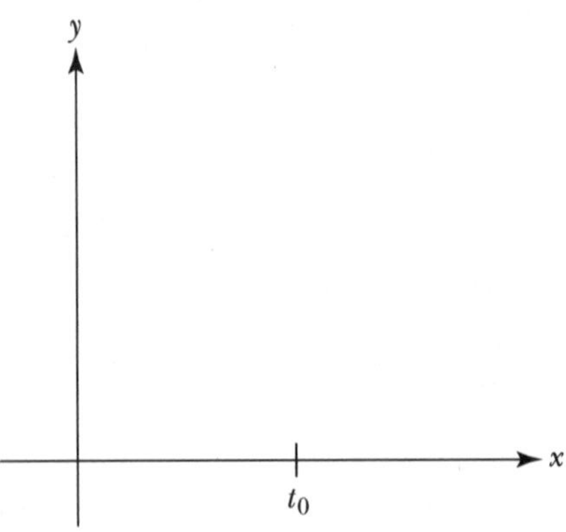

i. The tangent lines at points to the left and to the right of t_0 have positive slopes. The slope of the tangent line at t_0 is 0.

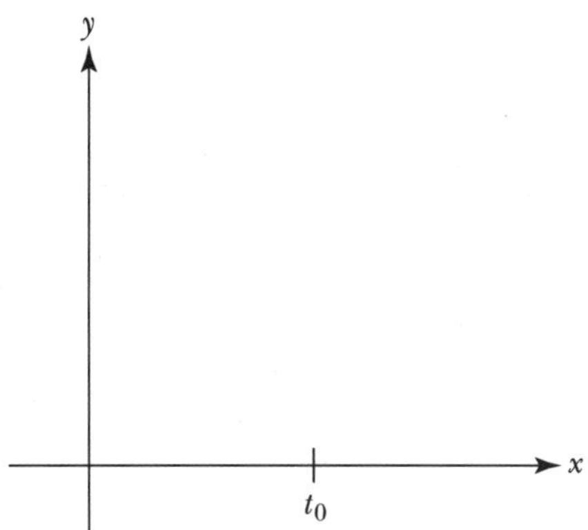

j. The tangent lines at points to the left and to the right of t_0 have negative slopes. The slope of the tangent line at t_0 is 0.

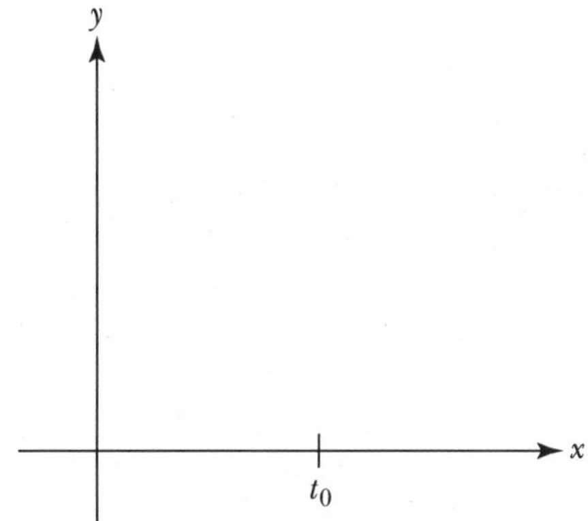

k. The function is increasing. The rate of change is increasing—that is, the slope of the tangent line is increasing as the line moves along the curve from left to right.

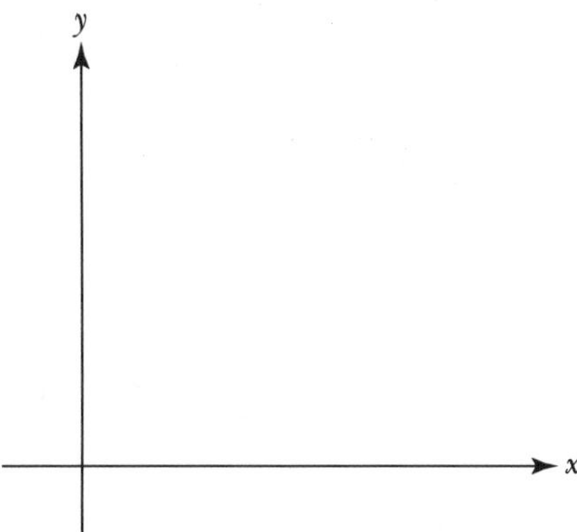

1. The function is increasing. The rate of change is decreasing—that is, the slope of the tangent line is decreasing as the line moves along the curve from left to right.

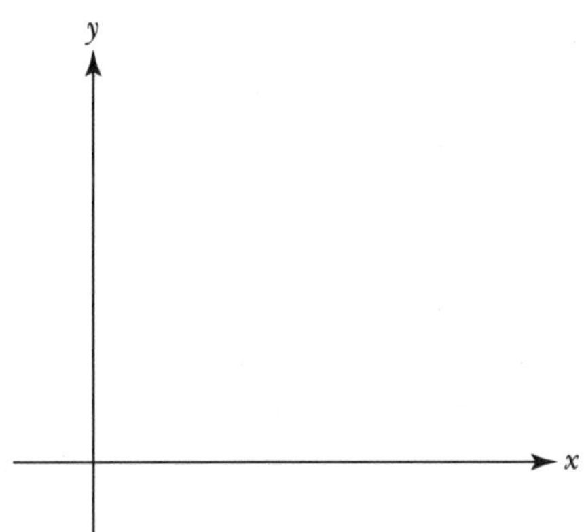

Whenever you use the motion detector to create a function, the graph lies in the first quadrant of the xy-plane, since, in this case, time and distance are positive quantities. This, of course, is not always what happens. For example, if you lived in Upstate New York and graphed the temperature (in Celsius) versus time over a one-year period, your graph would dip below the horizontal axis since there would be a number of days when the temperature drops below freezing. One way to indicate when the graph of a function lies below the horizontal axis, when it touches the axis and when it lies above the axis, is to use a *sign chart.* In the next task, you will create a sign chart for a given function. In addition, you will reverse the process and sketch the graph of a function which satisfies a given chart. But first a few words about functional notation.

When you represent a function by an expression, you give the independent variable a given name, such as x or *price*, and then define the dependent variable in terms of the independent variable. There are several ways to do this. One approach is to give the dependent variable a name, such as y or *tax*, and represent the function by equations such as

$$y = x^2 - 5x - 1 \qquad \text{or} \qquad tax = 0.06 \cdot price$$

Another approach is to give the function a name, such as f or T, in which case you would write

$$f(x) = x^2 - 5x - 1 \qquad \text{or} \qquad T(price) = 0.06 \cdot price$$

where $f(x)$ is read "f of x" and $T(price)$ is read "T of *price*."

To find the output value corresponding to a given input value, you assign the independent variable the desired value and solve for the value of the dependent variable. For example, when $x = 4$,

$$y = 4^2 - 5 \cdot 4 - 1 = -5 \qquad \text{or} \qquad f(4) = 4^2 - 5 \cdot 4 - 1 = -5$$

Consequently, the point $P(4,-5)$ lies on the graph of f. Moreover, since the value of $f(4)$ is negative, P lies below the horizontal axis.

The *sign of* $f(x)$ is the sign of the output value corresponding to x. For instance, the sign of f is negative $(-)$ when $x = 4$, since $f(4) = -5 < 0$. Similarly, the sign of f is positive $(+)$ when $x = 6$, since $f(6) = 5 > 0$. You can record this information on the sign chart for f as follows:

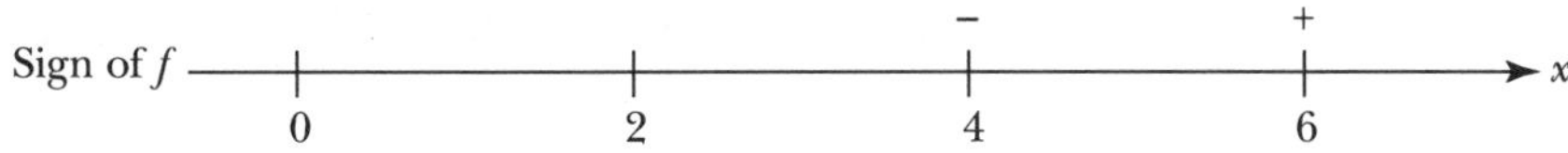

But $x = 4$ and $x = 6$ are only two of the values in the domain of f. What happens at the other values in the domain? What does the sign chart tell you about the location of the graph of f? You will explore answers to these questions in the next task.

Task 1-9: Interpreting Sign Charts

1. Create some sign charts.

a. Consider the function $f(x) = 2x - 2$.

(1) Sketch the graph of f on the pair of axes given below. Label the vertical axis.

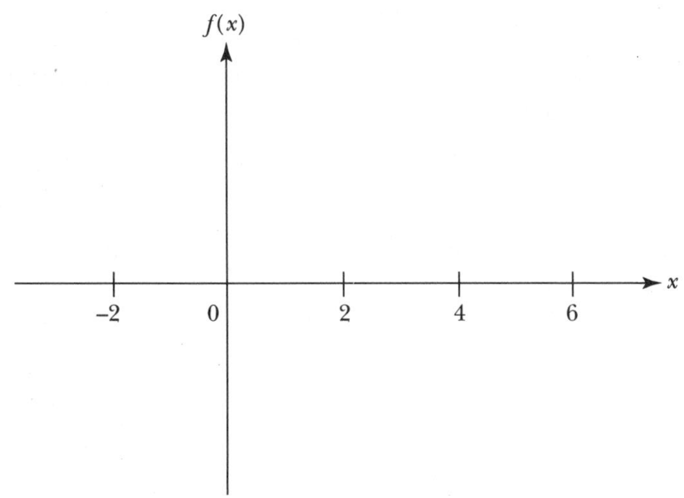

(2) Find all the values of x where $f(x) < 0$. In other words, find all the values of x where the graph of f dips below the horizontal axis.

(3) Find all the values of x where $f(x) = 0$. In other words, find all the values of x where the graph of f touches the horizontal axis.

(4) Find all the values of x where $f(x) > 0$. In other words, find all the values of x where the graph of f lies above the horizontal axis.

(5) Fill in the sign chart for f and indicate what the information on the sign chart says about the behavior of the graph of f by completing steps i and ii given below.

Note: We have used the same labeling for the horizontal axis in the graph of f and for the sign chart of f. Place the sign chart under the graph, and line up the labels. Doing this should help you see the connection between the chart and the graph.

Step i. Fill in the your observations about the sign of f. **Over** the axis for the sign chart:

- Indicate the intervals in the domain where $f(x) < 0$ by placing a string of "$-$"s over each interval.
- Indicate the values of x where $f(x) = 0$ by placing a "0" over each value.
- Indicate the intervals in the domain where $f(x) > 0$ by placing a string of "$+$"s over each interval.

Step ii. Fill in the information about (info re) the graph of f. **Beneath** the axis for the sign chart:

- Indicate the intervals in the domain where the graph of f dips below the horizontal axis by writing "below" under each interval.
- Indicate the values of x where the graph of f intersects the horizontal axis by writing "intersects" under each value.
- Indicate the intervals in the domain where the graph of f lies above the horizontal axis by writing "above" under each interval.

b. Consider the function $f(x) = x^2$.

(1) Sketch a graph of f on the pair of axes given below for $-3 \leq x \leq 3$. Label the axes.

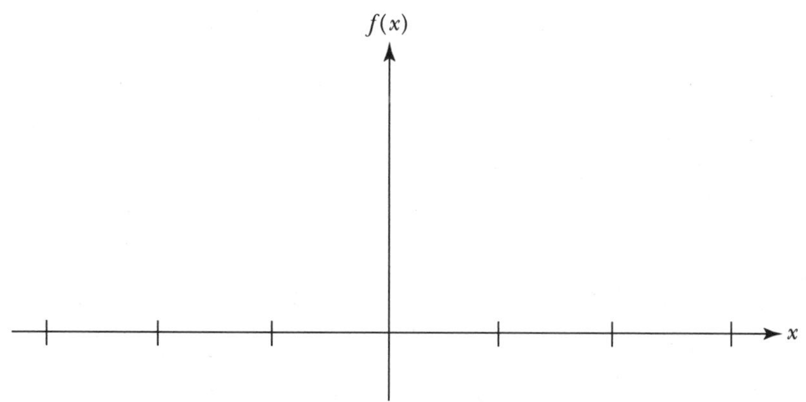

(2) Label the x values on the sign chart given below. Fill in the sign chart for f and indicate what the information on the sign chart says about the behavior of the graph of f.

Sign of f
Info re f ——— x

c. Consider the function

$$f(x) = \begin{cases} -1, & \text{if } x < -2 \\ x, & \text{if } -2 \leq x \leq 2 \\ 2, & \text{if } x > 2 \end{cases}$$

(1) Sketch a graph of f on the pair of axes given below for $-6 \le x \le 6$. Label the vertical axis.

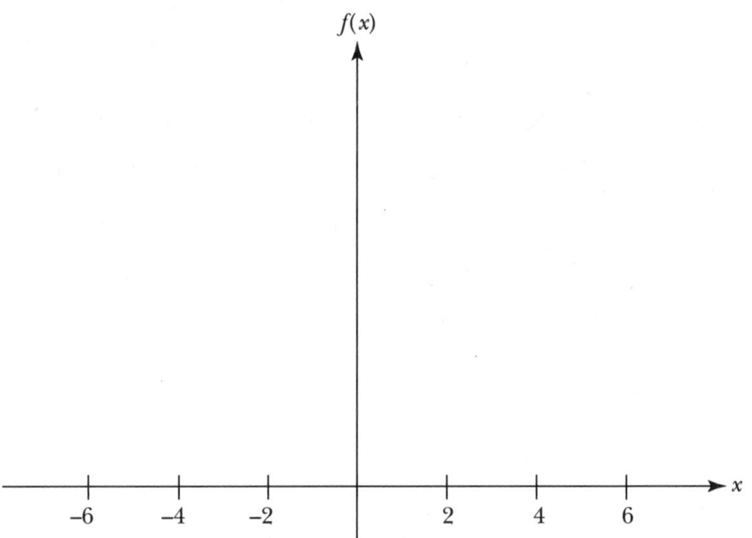

(2) Label the x values on the sign chart given below . Fill in the sign chart for f and indicate what the information on the sign chart says about the behavior of the graph of f.

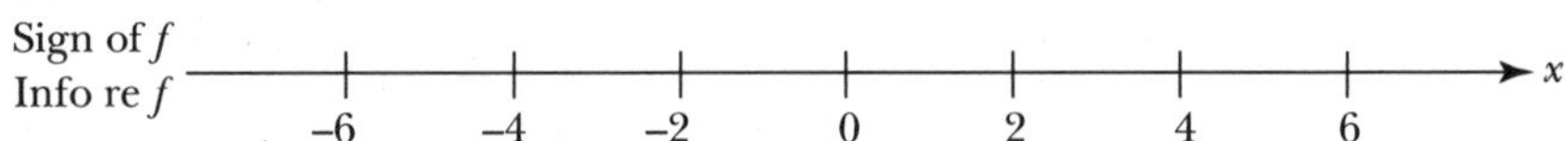

2. In the last part, you sketched the graph and then created the sign chart. In this part, reverse the process and use the chart (along with some other information) to sketch the graph.

Note: Before you sketch a graph, first indicate the information that the sign chart gives about the graph on the sign chart.

For each list of conditions given below, sketch a graph of a function satisfying all the conditions in the list.

a. The domain of f is all real numbers between -4 and 4.
f has a local minimum at $x = -2$.
f has local maxima at $x = 1/2$ and $x = -3$.

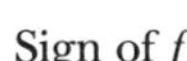

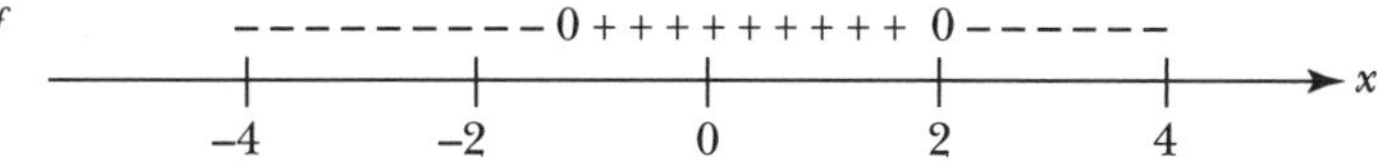

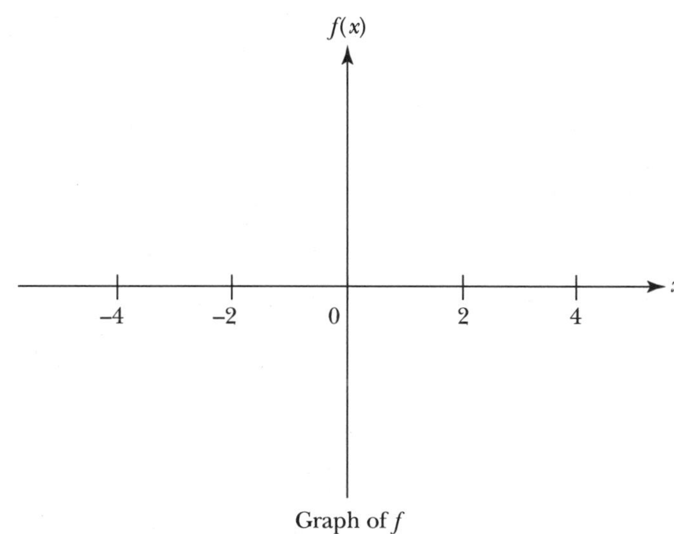

Graph of f

b. The domain of f is set of all real numbers.
f is increasing for $-\infty < x < -2$ and $2 < x < \infty$.
f is decreasing for $-2 < x < 2$.

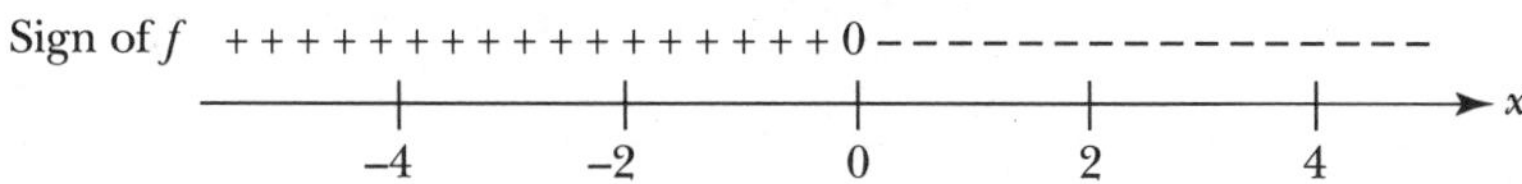

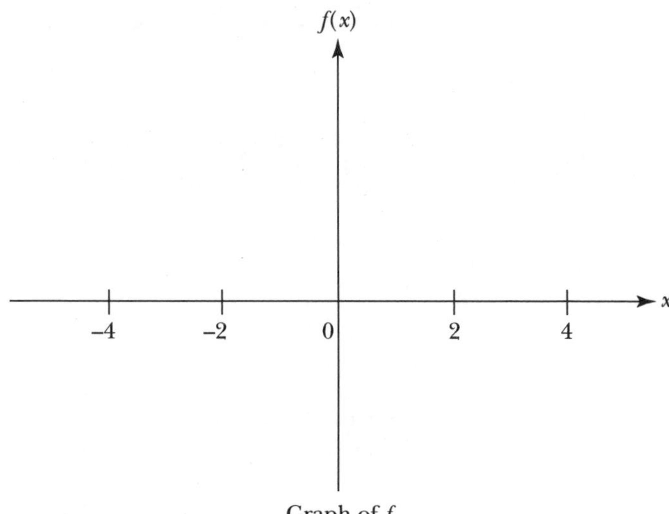

Graph of f

c. The domain of f is set of all x where $0 \leq x \leq 10$.

f has a turning point at each odd integer between 0 and 10.

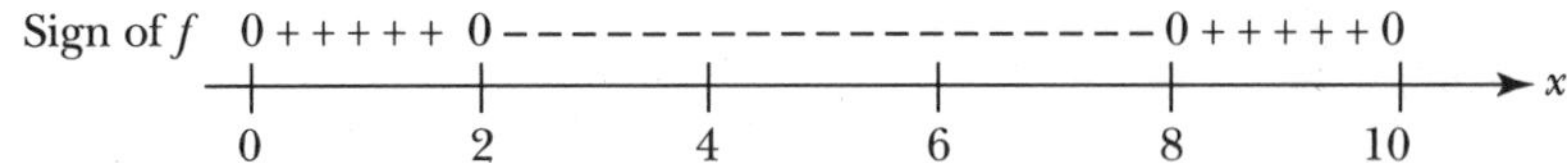

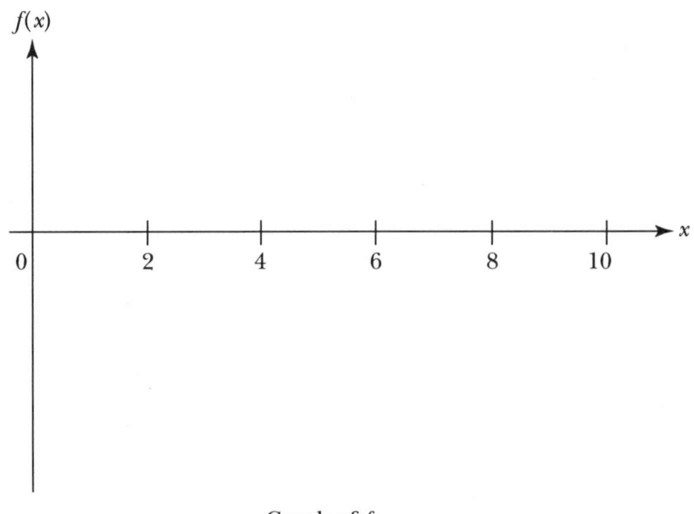

Graph of f

3. Summarize what is happening here by filling in the missing information in the table given below.

Sign of f	Graph of f	Sign chart
$f(x) < 0$	Below x-axis	—
$f(x) = 0$		
	Above x-axis	

Unit 1 Homework After Section 3

- Complete the tasks in Section 3. Be prepared to discuss them in class.
- Analyze some functions in HW1.9.

HW1.9 For each of the following functions:

i. Describe the behavior of the function. Label important points on the horizontal axis—*a*, *b*, *c*, and so on—and use these points to identify:

a. The intervals where the function is increasing.

b. The intervals where the function is decreasing.

c. The local maxima.

d. The local minima.

e. The intervals where the function is concave up.

f. The intervals where the function is concave down.

g. The points of inflection.

ii. Sketch the sign chart for the function.

1. **2.**

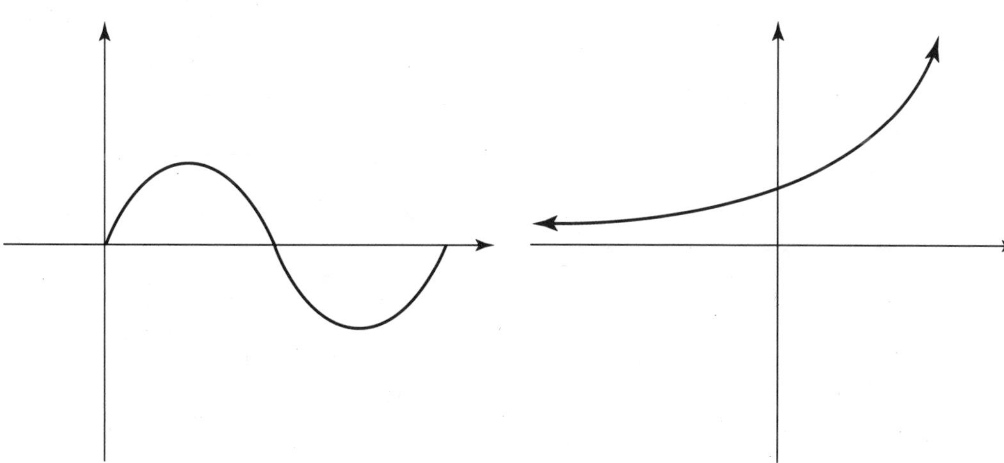

3. **4.**

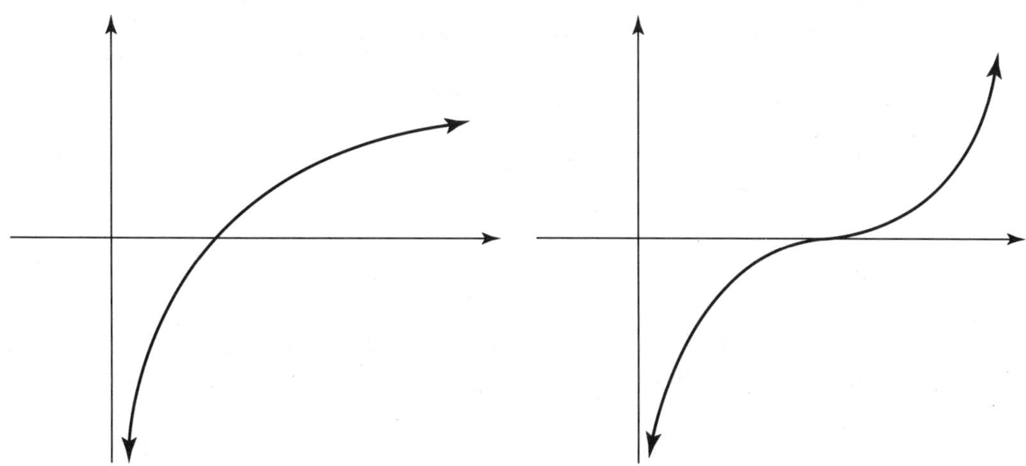

5.

6.

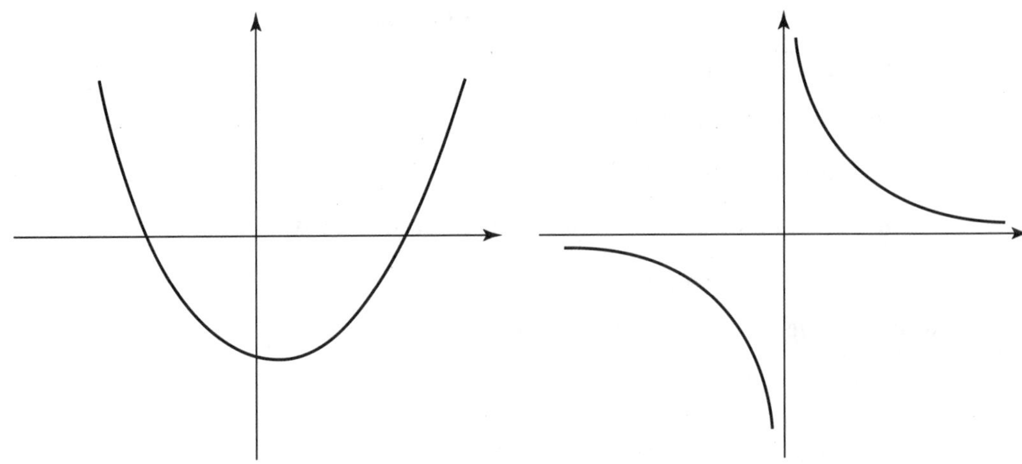

7.

8.

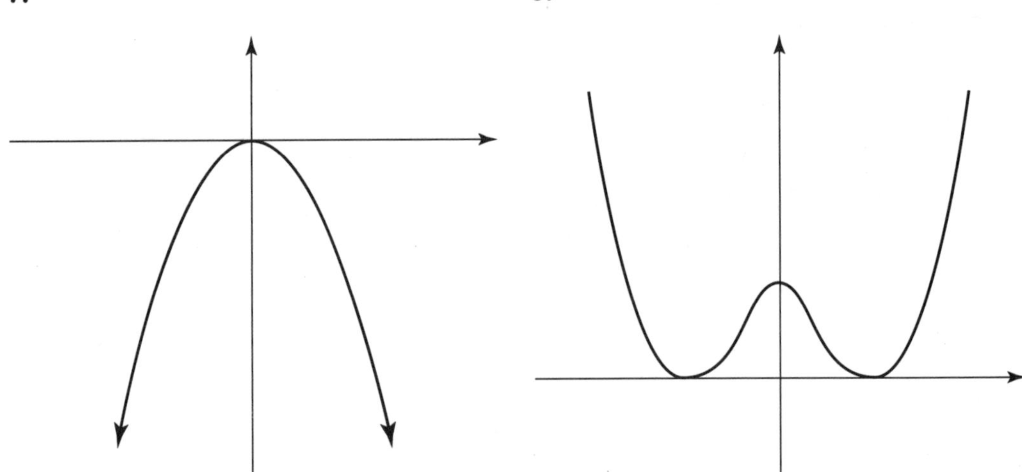

• Consider the definitions of increasing and decreasing in HW1.10.

HW1.10 Use inequalities to express the definitions of increasing and decreasing.

1. Sketch a curve with one turning point. Assume s is the name of the function and t is the name of the independent variable.

2. Consider the interval where s is increasing. Let t_1 and t_2 be any two points in this interval where $t_1 < t_2$. Mark t_1 and t_2 on the horizontal axis and the associated values of $s(t_1)$ and $s(t_2)$ on the vertical axis of your graph. Use an inequality to indicate the relationship between $s(t_1)$ and $s(t_2)$, when s is increasing.

3. Consider the interval where s is decreasing. Let t_3 and t_4 be any two points in this interval where $t_3 < t_4$. Use an inequality to indicate the relationship between $s(t_3)$ and $s(t_4)$, when s is decreasing.

- Show that you understand the concepts of domain, range, increasing, decreasing, concave up, and concave down by developing your own examples in HW1.11.

HW1.11 Develop examples that fit the following situations.

1. Draw a graph of a function whose domain is the set of all real numbers, which is increasing on the open intervals $(-4.5,0)$ and $(3,8.5)$ and decreasing everywhere else.
2. Give a table for a function whose domain is the set of integers from 1 to 10 and whose range is contained in the set of real numbers between 0 and 1.
3. Draw a graph of a function whose domain is the closed interval $[-5,5]$, whose range is the closed interval $[0,4]$, and which is always increasing.
4. Draw a graph of a function, whose domain is the open interval $(-5,5)$ whose range is the open interval $(0,4)$ and where the slope of the tangent line is always negative.
5. Draw a graph of a function whose domain and range are the set of real numbers and which has exactly two turning points.
6. Draw the graph of a function whose domain is the closed interval $[-10,8]$, which has a turning point at each odd integer in its domain, and has a local maximum at 0.
7. Draw a graph of a function whose domain is the open interval $(-30,30)$, where the slope of the tangent line to the graph at $x = -25, -15, -5, 5, 15$, and 25 equals 0.
8. Draw a graph of a function whose domain and range are the set of real numbers and where the graph is always concave up.
9. Draw a graph of a function whose domain and range are the positive real numbers and where the graph is concave up for all $x < 50$ and concave down for all $x > 50$.
10. Draw a graph of a function whose domain is the open interval $(-30,30)$, which is always increasing, and where the slope of the tangent line to the graph at $x = -25, -15, -5, 5, 15$, and 25 equals 0.
11. Draw a graph of a function whose domain is the open interval $(-10,8)$, whose range is the set of all real numbers, and which has local maxima at $x = -7, -4, 0, 5$, and 7.

12. Draw a graph of a function whose domain is the set of real numbers and where the slope of the tangent line is always 0.

- Think in terms of "rate of change" in HW1.12.

HW1.12

1. Consider the graph given below.

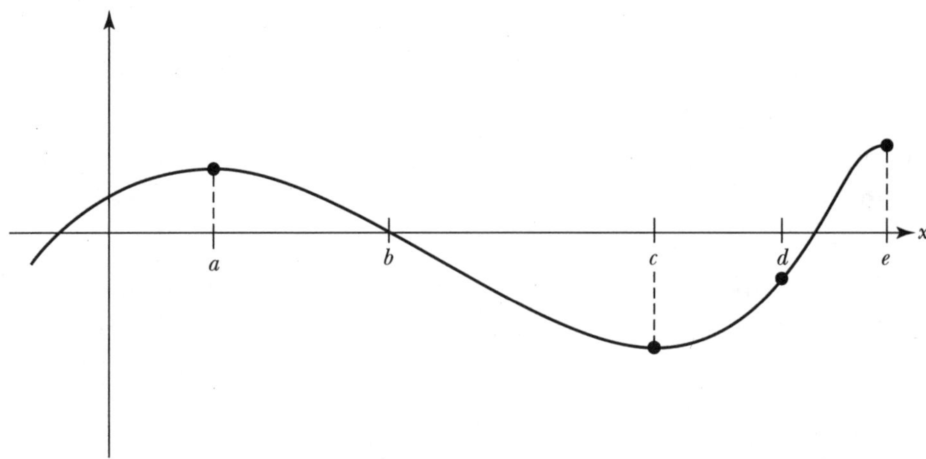

   a. At each of the labeled points, determine if the rate of change of the function—that is, the slope of the tangent line to the graph—is positive, negative, or zero.

   b. At which of the labeled points is the rate of change the greatest? Justify your response.

2. Consider the graphs modeling the growth of Plant #1 and Plant #2 which are described in the discussion preceding Task 1.7. Note: The scales for the graphs and the value of t_0 are the same.

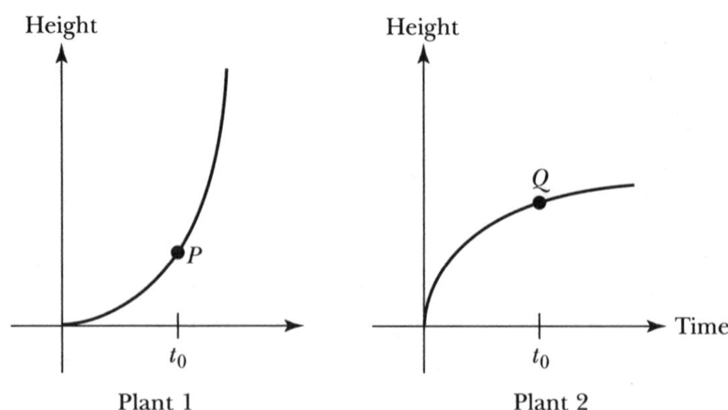

a. Describe the rate of growth—that is, the rate of change—of each plant by describing how the slope of the tangent line varies over the given time period.

b. Explain how you can tell by examining the graphs that at time t_0 Plant #1 is growing more rapidly than Plant #2.

c. Even though at time t_0 Plant #1 is growing at a faster rate than Plant #2, it is shorter than the other plant. Use a rate-of-growth argument to explain why this is the case.

d. At some point in time, Plant #1 catches up with Plant #2 and thereafter is taller than Plant #2. Indicate on the graphs the time when this change occurs.

- You cannot use the traditional two-point formula from algebra to find the slope of the tangent line to a smooth curve, since you only have one point, namely the point where the tangent line touches the curve. You can, however, approximate the value of the slope. Investigate how this might be done in HW1.13.

HW1.13 Approximate the slope of the tangent line by considering a nearby point on the curve.

The basic idea is this: Assume you have a function represented by an expression, and you have a point on the graph of the function, *P*. You want to calculate the slope of the tangent line to the graph at *P*. In order to be able to use the standard equation for finding the slope, namely

$$m = \frac{y_2 - y_1}{x_2 - x_1},$$

you need to know the xy-coordinates for two different points *on the tangent line.* Since *P* is on both the tangent line and the graph, you can find the xy-coordinates of *P* by substituting in the expression for the function. This gives you $P(x_1,y_1)$. The problem of course is that you don't know the coordinates of a second point on the tangent line. You can, however, use the expression for the function to find the xy-coordinates for a point *on the curve* which is *near P*, say $Q(x_2,y_2)$. The line through *P* and *Q* is called a *secant line.* Since you have two points, you can calculate the slope of secant line determined by P and Q. This provides you with an approximation of the slope of the tangent line at P.

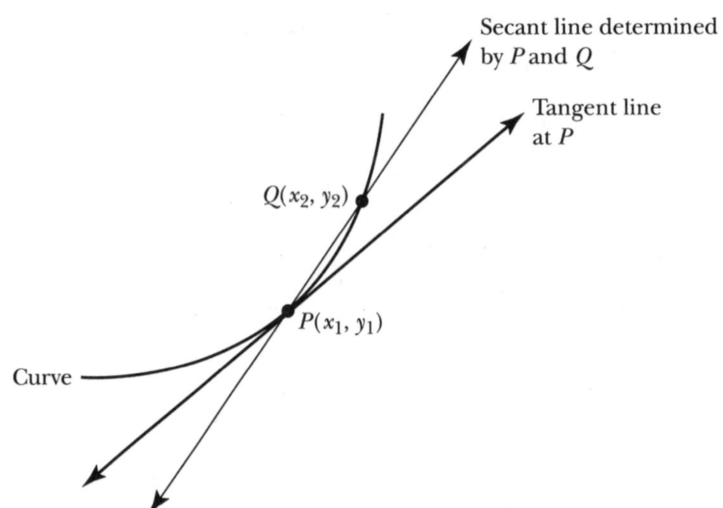

1. Before trying this approach on a specific example, study the diagram given above and make sure you understand the underlying idea.

2. Consider the parabola $y = x^2 + 1$. Use the secant line approach to approximate the slope of the tangent line—and hence the rate of change—of the function at the point $P(2,5)$.

   a. Sketch a large-scale graph of the parabola for $0 \leq x \leq 3$. Label the point $P(2,5)$ on the graph. Sketch the tangent line to the parabola at the point P.

   b. Choose a point Q on the parabola, which is near P. For instance, let $x = 2.25$. Use the expression for the parabola to find the second coordinate of Q. Label Q on the graph. Sketch the secant line determined by P and Q.

   c. Use the slope formula to find the slope of the secant line determined by P and Q.

   d. Your answer to part c gives you an approximation for the slope of the tangent line to the parabola at P. You can improve the approximation by considering a point Q which is closer to P, say $x = 2.1$. Try this.

3. Use the secant line approach to approximate the rate of change of the function $y = x^2 + 1$ at the point $P(-2,5)$.

- At the end of each unit you will be asked to write a journal entry reflecting on what you learned in the unit and on the learning environment for the course. Before you begin to write, review the material in the unit. Think about how it all fits together. Try to identify what, if anything, is still causing you trouble. Write your first journal entry in HW1.14.

HW1.14 Write your journal entry for Unit 1.

1. Reflect on what you have learned in this unit. Describe in your own words the concepts that you studied and what you learned about them. How do they fit together? What concepts were easy? Hard? What were the main (important) ideas? Give some examples of the main ideas.
2. Reflect on the learning environment for the course. Describe the aspects of this unit and the learning environment that helped you understand the concepts you studied. What activities did you like? Dislike?

Name ______________________________ Section __________ Date ______

Unit 2:

FUNCTION CONSTRUCTION

... I now refuse to teach ... unless a computer is available in the classroom. And I insist that students have access to computers outside of the classroom ... First, the powers of the computer can remove a great deal of drudgery. ... Secondly, the graphics capabilities of personal computers are wonderful.

John G. Kemeny 1988
Co-inventor of the BASIC Programming Language; Professor of Mathematics and Former President, Dartmouth College
In *Academic Computing* (May/June 1988), p. 59.

OBJECTIVES

1. Represent functions by:

   - verbal descriptions
   - expressions
   - piecewise definitions
   - graphs
   - scatter plots
   - sets of ordered pairs
   - tables

2. Convert between representations.

2

3. Distinguish among discrete, continuous, and almost continuous functions. Identify types of discontinuity.
4. Combine functions.
5. Reflect functions.

OVERVIEW

Most functions can be represented in several ways. You should think of the various representations as simply different ways of saying the same thing. For instance, a quadratic function can be represented by a graph (which is called a parabola), an expression [which has the form $f(x) = ax^2 + bx + c$, where $a \neq 0$], or a verbal description (which specifies the coordinates of its turning point and where the function intersects the horizontal axis).

Much of calculus involves analyzing one representation to get information about another—for instance, analyzing a messy expression to get precise information about its graph, without having any idea what the graph looks like ahead of time. Consequently, being able to convert between representations is an important part of being able to use calculus. For example, suppose you are a research scientist who studies food poisoning and is particularly interested in the impact of Salmonella bacteria. As part of your research, you have collected pages and pages of data giving the number of bacteria every 5 seconds over a 2-day period. Unfortunately, it is difficult to decipher what is happening by looking at the data represented by this huge table. However, by converting the table to a scatter plot it appears that the Salmonella are doubling every 20 minutes or so. To show that this is the case, you could represent the discrete data by an expression and then analyze the expression using calculus. In other words, you can use calculus to analyze the expression to get information about the table.

As you consider the various types of representations, you will consider ways to represent discrete functions (whose graphs consist of a collection of unconnected points) and piecewise continuous functions. You will develop an intuitive understanding of what it means for a function to be continuous and use observation to categorize some types of discontinuity, such as "jumps" and "holes."

You will also explore ways to create new functions from old ones. Just as you can combine numbers by adding, subtracting, multiplying, and dividing them to get a new number, you can combine functions to define a new function. This is especially useful if you need to model a complex situation which is comprised of individual components. Instead of considering the entire process, you can model the pieces and then combine them in an appropriate way to model the whole situation.

Finally, you will create new functions by reflecting old functions through various lines. For instance, if you have a parabola that has a local minimum at the origin, what does its graph look like when you flip it over, or reflect it through, the horizontal axis? How is the expression for the new function related to the expression for the original one? What happens if you flip the given parabola over the vertical axis? You will also explore how to reverse, or invert, the process of a function. For instance, with the Salmonella data, the function inputs the time and outputs the number of Salmonella at that time; if you invert the process, the function would input the number of Salmonella and output the corresponding time. What type of reflection do you need to make to reverse a process? Under what conditions is the inverse of a function also a function?

In Unit 1, you used the motion detector to create functions and develop an intuitive understanding about their behavior. In this unit you will use the programming language ISETL (Interactive SET Language) to construct functions. ISETL was developed as a tool for helping students acquire a firm understanding of basic mathematical concepts, such as functions, sets, and logic. It is an *interactive* language; when you give ISETL some information, it processes the information and gives a response. Thinking about how ISETL might do this will help you form mental images associated with fundamental mathematical ideas, such as the process of a function or the behavior of a limit. ISETL's syntax is very similar to standard mathematical notation—when you write something in ISETL, it looks like mathematics, and vice versa. Writing ISETL will help you learn to write precise mathematical notation, since the computer needs to have its input "just right"—there is no room for fudging.

SECTION 1

Representing Functions

Functions can be constructed using a number of different *representations*, including expressions, piecewise definitions, tables, sets of ordered pairs, graphs, and scatter plots. In this section you will use the programming language ISETL to construct functions using various representations. For each representation, you will examine how the domain of the function is specified and how the action of the function is implemented. You will convert between the various representations.

As you read and write short ISETL code segments, "Think ISETL." That is, after you type in a line of code, but before you press the return key:

1. Think about how ISETL might process the information that you have given it and predict what output ISETL is going to give you.
2. Write down your prediction.
3. Press the return key.
4. Compare your prediction to ISETL's actual output.

In other words,

Type—Predict—Return—Compare

Try not to skip this step. Thinking ISETL will help you to:

- envision a function taking an input object and transforming it into a corresponding output object, and
- construct new functions.

ISETL is a powerful mathematical programming language with many different features. You are only going to learn a few. In the next task, you will examine how to evaluate an expression in ISETL, assign a value to a variable, and use some predefined ISETL constants and functions. Experiment. Play. Try to figure out what is going on. Think ISETL! Look carefully at the code which is given to you. When the time comes for you to write your own code, mimic examples you have seen and use your course handout on ISETL.

Task 2-1: Getting Started with ISETL

1. Read the first few sections in your course handout on ISETL.
   - Learn how to enter and exit ISETL.
   - Note the need for a semicolon at the end of a line.
   - Note the difference between a single (>) and a double (>>) prompt.
   - Note the difference between an expression (which returns a value) and a command (which tells ISETL to do something, but does not return a value).
   - Review the discussion of arithmetic, Boolean and relational operators, predefined constants and functions, and assigning a value to a variable.
   - Learn how to edit and re-execute code fragments.

   Your course handout on ISETL is a valuable resource. Turn to it when you have questions. Become familiar with it.

2. Open the ISETL program. Enter the following arithmetic expressions and see what happens. In each case try to predict what ISETL is going to return *before* pressing the return key. That is, Think ISETL!

ISETL Code	Predicted Output	Actual Output
2.01 + 3.24;		
2*(6 − 7)/4;		
3**2;		
23.879 − 1.45 ;		
2**(3**2);		

3. Assign a value to a variable. Evaluate some expressions containing Boolean and relational operators.

Enter the following code. As usual, predict what ISETL will output before you press the return key. If there isn't any, then write "NO" for "No Output." If you make an error, edit your code.

ISETL Code	Predicted Output	Actual Output
x := 2;		
x**4;		
y := 6.0;		
y;		
x + y;		
y**x;		
z;		
z := −35.4;		
z;		
z >= 0.0;		
z /= −100;		
z < −35;		
(−50 <= z) and (z <= −30);		
(z < −40.0) or (z > 0);		

4. Use some of the predefined ISETL constants and functions. See if you can figure out what each function does.

ISETL Code	Predicted Output	Actual Output
x := 16;		
sqrt (x);		
sqrt (−x + 10);		
abs (−45.8);		
abs (3*x);		
sin (0);		
t := Pi;		
sin (t/2);		

The next task will introduce you to the ISETL syntax for representing functions defined by one or more expressions. First you will enter some code and think about what's going on, and then you will write your own code. As you examine each of the following representations, ask yourself:

- How is the domain specified in this representation?
- How can you determine the range?
- How does the representation implement the process of the function? That is, how is the relationship between an input and its associated output indicated?

Task 2-2: Implementing Functions Using ISETL funcs

1. Represent a function defined by an expression in ISETL.

   a. Enter the following code in ISETL.

      Pay careful attention to the syntax for an ISETL **func**, so that you will be able to write your own code in the next part of this task.

Each time you ask ISETL to evaluate a function at a particular input, think about how the function (which is represented by an ISETL **func**) will process the input. Trace through the code for the function and predict the associated output. Take your time. Think about what's happening before you press the return key. Form an image in your mind of the function processing the information.

ISETL Code	Predicted Output	Actual Output

f := func(x);
    if x >= 0 then
        return x**2;
    end if;
end func;		

f(4);		
f(−5.57);		
x := 6;		
f(x);		
x/2;		
f(x/2);		
r := 8.0;		
f(r);		
f(2*r + 4);		
f(−r);		
g := func(x); return 3*x + 8; end func;		
g(10);		
f(10);		
f(10) + g(10);		
2*f(10) − g(10);		
g(4) / f(4);		

b. Answer some questions about the function f defined in the previous ISETL session.

(1) Give the domain of f.

(2) Give a verbal description of the process f uses to transform an input item into a new item.

(3) Give a mathematical description of f—that is, represent f by an expression.

(4) Sketch a graph of f. Label your axes carefully.

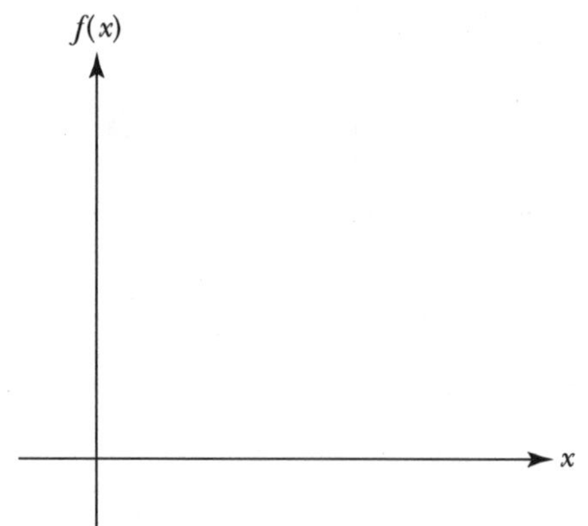

(5) Give the range of f.

c. Review the syntax for using a **func** to construct a function represented by an expression which is contained in your course handout on ISETL. In the space below, give the general ISETL syntax for this case.

d. Construct some functions.

Enter the code for each of the following **funcs** in ISETL. Save the code for the **funcs** in a file. Print a copy of the file. Place the code for each **func** in the designated space.

(1) $h(t) = \frac{1}{t}$, where $t \neq 0$.

(a) Fill in the following table by evaluating h (by hand or using your calculator). Express your results in decimal form. If $h(t)$ does not exist, write **om**—or undefined—in the table.

t	−10	−1/10	0	0.1	1/2	5	1,000
h(*t*)							

(b) Represent h by an ISETL **func**. Enter the code for your **func** in ISETL. Use your **func** to check your answers to part (a). Place the code for your **func** below.

(2) $g(r) = \sqrt{r - 1}$, where $r \geq 1$.

(a) Evaluate g (by hand) at the specified input values. If $g(r)$ is undefined, write **om** in the table.

r	0	1	65	100	1.25	101/100
$g(r)$						

(b) Represent g by an ISETL **func** using the predefined ISETL function **sqrt**. Use your **func** to check your answers to part (a). Place the code for your **func** below.

(3) f accepts any negative real number as input. f cubes the input value, changes the sign of the result, and then returns its reciprocal.

(a) Give the domain of f.

(b) Give a mathematical description of f.

(c) Evaluate f at the given input values.

x	−10	−1	−1/2	−0.01	0	10.7
$f(x)$						

(d) Represent f by an ISETL **func**. Use your **func** to check your answers to part (c). Place the code for your **func** below.

2

(4)

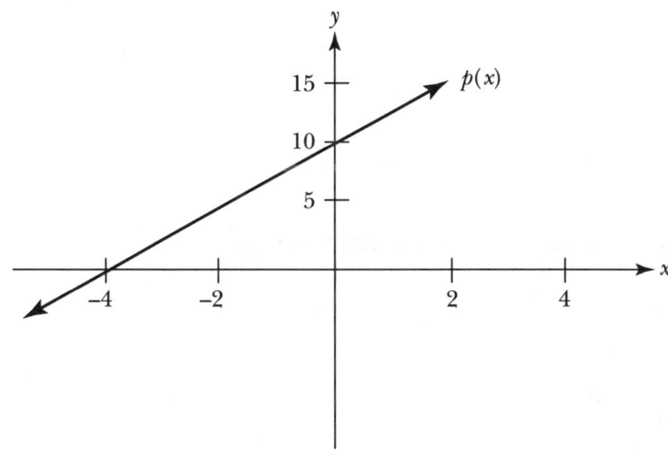

Assume the domain of p is the set of all real numbers.

(a) Give a mathematical description of the function p.

(b) Evaluate p at the given input values.

x	−6	−4	0	1/4	10	40
p(x)						

(c) Represent p by an ISETL **func**. Omit the **if ... then ... end if** portion of the **func** since there are no restrictions on the domain. Use your **func** to check your answers to part (b). Place the code for your **func** below.

2. Represent a piecewise-defined function using an ISETL **func** and an **if ... then ... elseif** ... statement.

   a. Enter the following code in ISETL. As usual, pay careful attention to the syntax. Whenever you instruct ISETL to evaluate a function at an input value, predict ISETL's output by tracing through the code defining the function before pressing the return key.

ISETL Code	Predicted Output	Actual Output

g := func(t);
    if (−10 < t) and (t < 0) then
      return −t + 1;
    elseif t = 0 then
      return 5;
    elseif t > 0 then
      return t − 1;
    end if;
end func;		

g(−15.6);		
g(−2.7);		
g(0);		
x := 2.0;		
g(x);		
g(x**2 + 5);		
h := func (x); return 2 * (x**3); end func;		

h(4);		
g(h(4));		
g(6.0);		
h(g(6.0));		
6 * (g(0) + h(0));		
(3 * h(−1) − 10 * g(−1)) / 13;		

b. Answer some questions about the piecewise-defined function g defined in the previous ISETL session.

(1) Give a verbal description of the process g uses to transform an input object into a new item.

(2) Give a mathematical description of g by representing g by a piecewise-defined function using the curly bracket notation.

(3) Sketch a graph of g. Label your axes carefully.

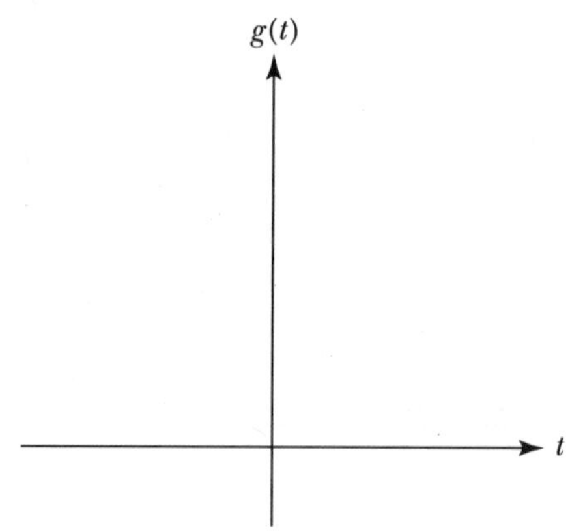

(4) Give the domain of g.

(5) Give the range of g.

c. Review the syntax for constructing a piecewise-defined function using a **func** which is contained in your course handout on ISETL. In the space below, give the general ISETL syntax for representing a piecewise-defined function.

d. Represent some piecewise-defined functions using ISETL **funcs** that contains an **if** . . . **then** . . . **elseif** . . . statement.

Enter the code for each **func** in ISETL. Save the code for the **funcs** in a file. Print a copy of the file. Place the code for each **func** in the designated space.

(1)

$$f(t) = \begin{cases} t^2, & \text{if } t < 0 \\ -t, & \text{if } 0 \le t < 5 \\ -2, & \text{if } t \ge 5 \end{cases}$$

(a) Evaluate f (by hand) at the given values.

t	−20	−0.25	0	3.75	5	40
$f(t)$						

(b) Represent f by an ISETL **func**. Use your **func** to check your answers to part (a). Place the code for your **func** below.

(2) If an input value is negative, the function finds its reciprocal; otherwise it returns its square root.

(a) Give the domain of the function.

(b) Give a mathematical description of the function.

(c) Evaluate the function at the input values in the table.

x	−4	−0.25	0	0.25	9	11.8
$f(x)$						

(d) Represent the function by an ISETL **func**. Use your **func** to check your answers to part (c). Place the code for your **func** below.

(3)

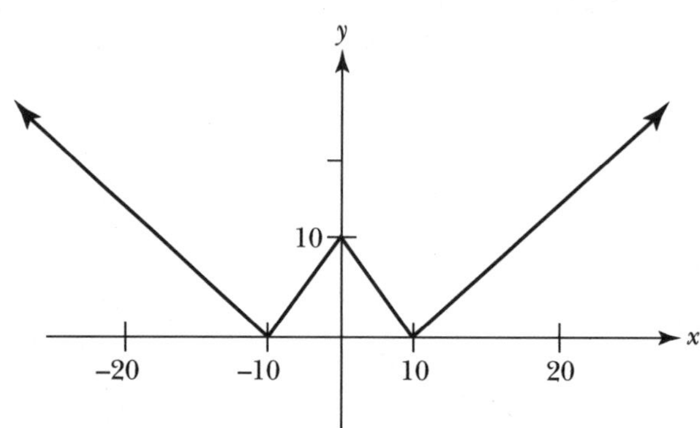

(a) Assume the domain of the function is the set of all real numbers. Give a mathematical description of the function.

(b) Evaluate the function at the given input values.

x	−20	−10	−5	0	5	10	20
f(x)							

(c) Represent the function by an ISETL **func**. Use your **func** to check your answers to part (b). Paste the code for your **func** below.

Thus far, you have examined two ways to construct functions in ISETL. First, you investigated how a function represented by an expression can be constructed using a **func**. In this case, the **func** accepts an input value and tests if it's in the domain. If it is, the **func** implements the process of the function by substituting the value into the expression, performing the indicated computations, and returning the result. If the value is not in the domain—in which case, ISETL reaches the end of a **func** without encountering a **return** statement—then the **func** returns **om**.

You also investigated how a *piecewise-defined* function can be represented using an ISETL **func** containing an **if...then...elseif...** statement. In this case, the **func** accepts a value of the independent variable, repeatedly tests to determine if it's in a particular subdomain, and if it is, evaluates the associated expression and returns the result.

The domains of the functions which you considered in the last task contained an infinite number of points, such as the set of all real numbers or the set of all numbers between −3 and 10. The graphs of these functions consisted of lines and curves. The graph of a *discrete* function, on the other hand, consists of a collection of distinct, unconnected points. A discrete function whose domain contains a finite number of values can be represented by a table, a set of ordered pairs, or a scatter plot. In the next two tasks, you will investigate how to use ISETL to construct discrete functions and how to convert back and forth between ISETL representations and mathematical representations.

A few words about some mathematical terminology before you begin: A *set* is a collection of objects which has no order. It does not make any sense to talk about the position or location of an item in a set. A *sequence*, on the other hand, is an ordered collection of objects. There is a first item, a second item, and so on. Sequences are represented by **tuples** in ISETL.

Task 2-3: Constructing Discrete Functions Using Sets of 2-tuples in ISETL

Represent a discrete function defined by a set of ordered pairs or a table in ISETL.

1. Enter the following code in ISETL. "Think ISETL!"

ISETL Code	Predicted Output	Actual Output
h := {[5, 22], [4,11], [−3, 13], [0,5], [1,7], [−1,4], [−5,26]};		
h(−3);		
h(4);		
h(11);		
t := 5;		
h(t);		
h(t − 6);		
m := {[1,2], [4,−5], [11,4], [7,1], [−3,5]};		
h(4) * m(4);		
h(m(4));		
x:= 7;		
h(m(x));		
m(h(−5));		

2. Answer some questions about the function h defined in the previous ISETL session.

a. Give the domain of h.

b. Give the range of h.

c. Explain how to find the output associated with a given input.

d. Represent h by a graph which is called a *scatter plot.*

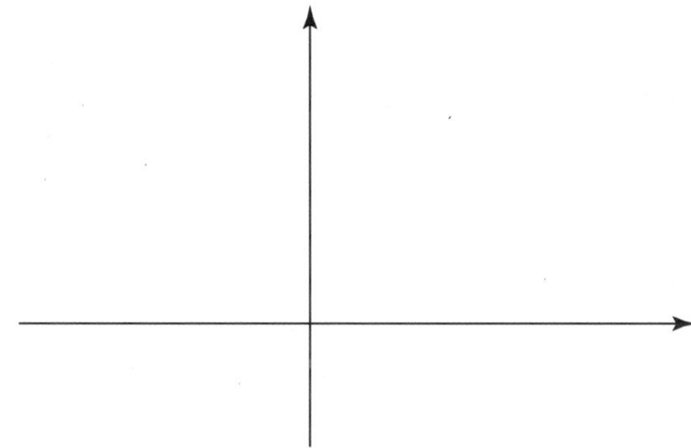

2

3. Give the general syntax for constructing a function defined by a set of ordered pairs in ISETL.

4. Represent some discrete functions in ISETL using sets of 2-tuples. Give the code for each function.

   a. The domain of the function is the odd integers between -11 and -1 (inclusive). The function returns the square of each input value.

      (1) Give the domain of the function.

      (2) Give the range of the function.

      (3) Represent the function in ISETL as a set of 2-tuples.

b.

x	−2.5	−2.0	−1.5	−1.0	−0.5	0	0.5
y	3.4	−4.8	0	3.4	5.6	0	7.1

(1) Give the domain of the function.

(2) Give the range of the function.

(3) Represent the function in ISETL as a set of 2-tuples.

c.

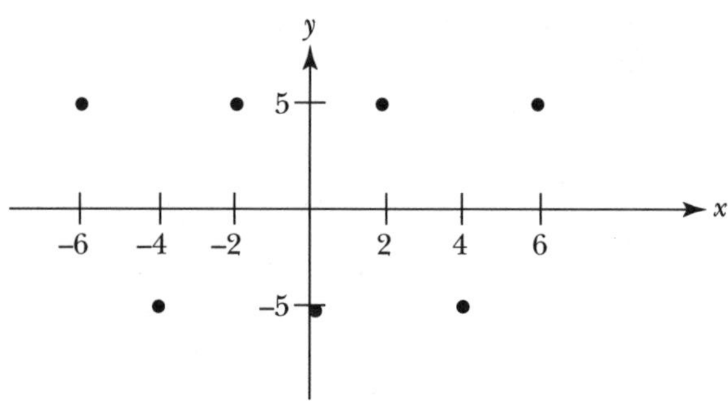

(1) Give the domain of the function.

(2) Give the range of the function.

(3) Represent the function in ISETL as a set of 2-tuples.

A discrete function that is represented by a *table* or *set of ordered pairs* can be implemented in ISETL by a **set of 2-tuples**. Each *xy*-entry in the table or each ordered pair in the set is represented by an ISETL 2-tuple. The set of all first components of the 2-tuples determines the domain of the function. The set of all second components determines the range. The process of the function is implemented by assigning the second value in the 2-tuple to the first. Sometimes you can see a pattern indicating how the function is defined, and sometimes you can't.

A discrete function can also be implemented in ISETL using a graph, which is called a scatter plot since it consists of a bunch of scattered or unconnected points. A function which is represented by an expression or defined piecewise can also be implemented in ISETL using a graph. Although in this case most of the points on the graph are connected, it is interesting to note that ISETL does not connect them. Why is this? A computer or calculator determines the shape of a graph by sampling a finite number of points in the domain and evaluating the function at these points. It cannot possibly consider every point in the domain since it is a finite—discrete—machine. The question then is: Do the points which the computer has chosen provide an accurate reflection of the behavior of the function? Unfortunately, a finite sample cannot be used to predict with absolute certainty what happens throughout the domain. Connecting points in a finite sample may connect points that should not be connected, thereby giving a misleading picture of what is really going on. It is possible, for instance, that between two sample values the function "blows up," has a "hole," or oscillates rapidly. These types of behaviors may be missed. Consequently, the ISETL **plot** command shows you what happens at the points in the finite sample. It does not jump to any conclusions about what happens in between these values.

Task 2-4: Representing Functions by ISETL Graphs

1. Read about the ISETL **plot** command in your course handout on ISETL.

   - Learn how to enter and quit the Graphics Window.
   - Review how to specify the domain and range of a function.
   - Review how to find the coordinates of a point and zoom in on a region.

2. Enter the following code in ISETL.

   If the output is a graph, predict the shape of ISETL's graph by drawing a sketch. Compare your sketch to ISETL's. If they differ, explain why.

   Experiment. Change the bounds on the domain. Change the bounds on the range. See what happens.

Caution: In a computer representation of a graph, the horizontal and vertical axes may cross at a point other than the origin. When you examine a graph, always observe where the axes intersect.

ISETL Code	Predicted Graph

f := **func** (t);
  if t <= 0 then
    return sqrt (−t);
  else
    return sqrt (t);
  end if;
end **func**;	

plot (f);	
plot (f, −25,0);	
plot (f, −10, 100, 0,10);	

```
s := {[0,0], [0.5,4], [2.5,-1.5], [4,-2.5], [6,2], [7.5,5], [9,-1]};
```

```
plot (s, 0, 10);
```


2. Answer some questions about representing a function by a graph.

a. Describe how to determine the domain of a function by looking at its graphical representation.

b. Describe how to determine the range of a function by looking at its graphical representation.

c. Explain how you use a graph to find the output that corresponds to a given input.

3. Represent some functions using ISETL graphs. Analyze the behavior of the functions by observing their graphs.

   a. $f(x) = \sin(x)$, where $-6.28 \leq x \leq 6.28$.

      (1) Give the domain of f.

      (2) Represent f by an ISETL **func** using the predefined ISETL function **sin**. Write the code for the function below.

      (3) Represent f using an ISETL graph. Be sure and specify the endpoints of the domain. Sketch the graph below.

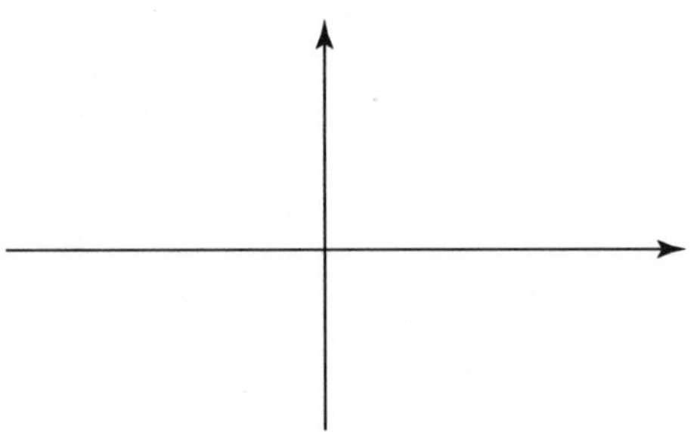

      (4) Use the ISETL graph to approximate the range of f.

      (5) Use the ISETL graph to analyze the behavior of f. Approximate:

         (a) The intervals where f is increasing.

(b) The intervals where f is decreasing.

(c) The coordinates of the local maxima of f.

(d) The coordinates of the local minima of f.

(e) The intervals where the graph of f is concave up.

(f) The intervals where the graph of f is concave down.

b. $g(r) = 3r$, where $r = -2, -1.5, -0.5, 0, 0.5, 1, 1.5, 2$.

(1) Represent g by a mathematical table.

(2) Give the domain of g.

(3) Give the range of g.

(4) Implement *g* in ISETL by a set of 2-tuples. Write the ISETL code for *g* below.

(5) Represent *g* by an ISETL scatter plot. Sketch the graph below.

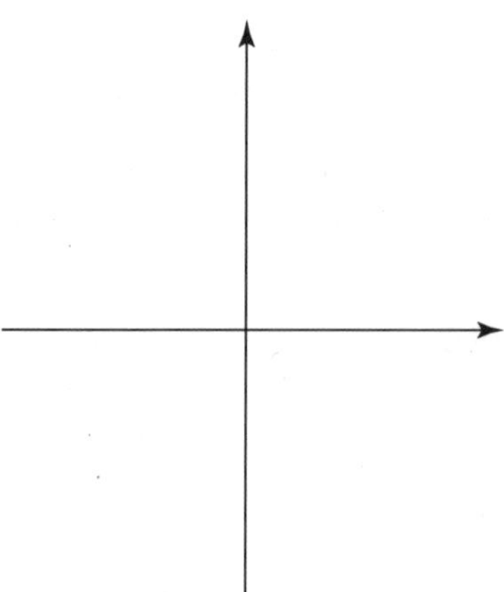

In conclusion, a *graph* or *scatter plot* can be represented using the ISETL command **plot**. The domain is determined by drawing a vertical line from each point on the graph to the horizontal axis. The process of the function is implemented by starting at a value in the domain, traveling in a vertical direction until you reach the graph, and then traveling in a horizontal direction to the corresponding value on the vertical axis.

In the homework problems you will have an opportunity to implement some more functions in ISETL and use ISETL to help analyze their behavior.

Unit 2 Homework After Section 1

- Complete the tasks in Section One in the Activity Guide. Be prepared to discuss them in class.
- Review the syntax for representing functions in ISETL.
- Implement some functions in ISETL and analyze their behavior in HW2.1.

HW2.1 Use functions to model some situations.

1. According to Podunk's income tax schedule, a person who earns less than \$10,000 does not have to pay any taxes. A person who earns at least

$10,000 but less than $50,000 must pay 15% of the amount he or she earns in taxes. Anyone who earns $50,000 or more must pay $6,000 plus 27% of the amount earned over $50,000. Model Podunk's tax schedule, and use the model to determine if the schedule is fair.

a. Calculate the tax bill for a person who earned:
(1) $6,798
(2) $23,456
(3) $74,892

b. Suppose *T* is a function that accepts the amount of money an individual earns and returns the person's tax bill. Give a mathematical description of *T*. State the domain of *T*.

c. Represent *T* by an ISETL **func**. Give the code for your **func**. Use your **func** to check your answers to part a.

d. Represent *T* by a graph, using the ISETL **plot** command. Sketch the graph.

e. Find the range of *T*.

f. Is the tax scheme fair? Justify your response.

2. "The Pole-Vault Principle: If every pole-vaulter's technique and strength were equal, judges could skip the actual vault and determine the height an athlete would reach by clocking his speed at the point he plants the pole. The equation is straightforward: height equals the square of the velocity divided by twice the gravitational constant (9.8 m/sec^2). But given a particular running speed, explains NYU's Richard Brandt, a taller vaulter has a bit of an edge because his center of mass is higher and so requires a bit less energy to reach a particular height. Lesson: Be fast; be tall; hang on." (*Newsweek*, 27 July 1992, p. 59)

Assume every pole-vaulter's technique and strength are equal and their velocity varies from 0 to 15 meters per second (m/sec).

a. Find a mathematical expression which determines the height an athlete at a particular velocity. State the domain of the function.

b. Find the height an athlete would reach if his or her velocity is:
(1) 8 m/sec
(2) 9 m/sec
(3) 10 m/sec

c. Represent the function which determines the height an athlete would reach at a particular speed using a **func**. Give the code for your **func**. Use the **func** to test your responses to part b.

d. Find the range of the function.

e. Represent the function by an ISETL graph. Sketch the graph.

3. Your uncle sells cold beer at the Baltimore Orioles' ballpark, Camden Yards. Based on his past experience (and very careful record-keeping), he knows that at a Saturday game, for each degree the temperature increases he sells eight more liters of beer than when it's 62°F. He sells 354 liters when it's 62°F, and the temperature varies between 62°F and 103°F.

   a. Calculate the number of liters he sells when the temperature is:
      (1) 65°F
      (2) 82°F
      (3) 97°F

   b. Find a mathematical expression that determines the number of liters of beer he sells for a given temperature. State the domain of the function.

   c. Represent your uncle's situation by an ISETL **func** called NumLiters. Give the code for your **func**. Use your **func** to check your answers to part a.

   d. Represent your function by an ISETL graph. Sketch the graph.

   e. Find the range of the function.

4. Carl has all sorts of problems learning to use the computer. In the beginning, everything that can go wrong goes wrong. But Carl never gives up on anything, so he hangs in there. The following is a graph of his learning curve, giving the rate of his errors per hour versus time in hours.

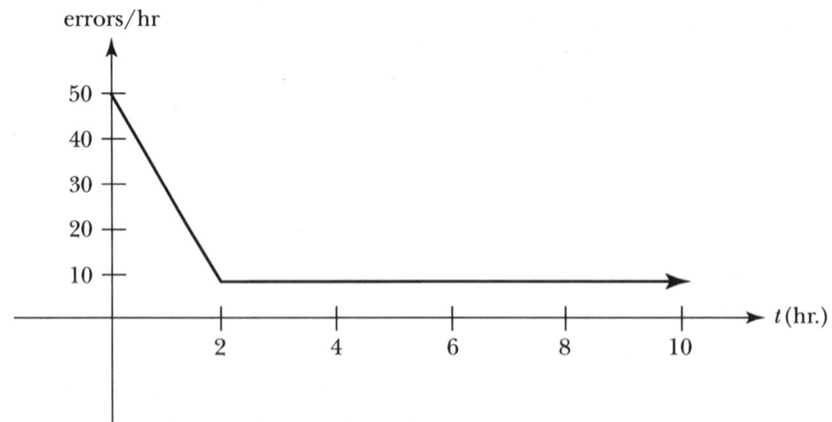

a. Find Carl's errors per hour at
  (1) 30 minutes
  (2) $2\frac{1}{2}$ hours
  (3) 75 minutes

b. Find the difference between Carl's error rate at 1 hour and his error rate at 5 hours.

c. Describe Carl's improvement as time goes by.

d. Model Carl's error rate at a given time using a piecewise-defined function. State the domain of the function.

e. Find the range of the function.

f. Represent the function by a **func**. Give the code for your **func**. Use your **func** to check your answers to part a.

- As you know, a discrete function whose domain and range are both finite sets can be represented by a table, set of ordered pairs and a scatter plot. It can also be represented by an "arrow diagram."

HW2.2 The general idea underlying an arrow diagram is to sketch two closed, wiggly regions, one corresponding to the domain of the function and the other containing the range. The process of a discrete function can be implemented by drawing an arrow from each point in the domain to its corresponding value in the range. For example, a function whose domain is {1,2,3,4} and whose range is contained in the set $\{a,b,c,d,e,f\}$ can be represented as follows:

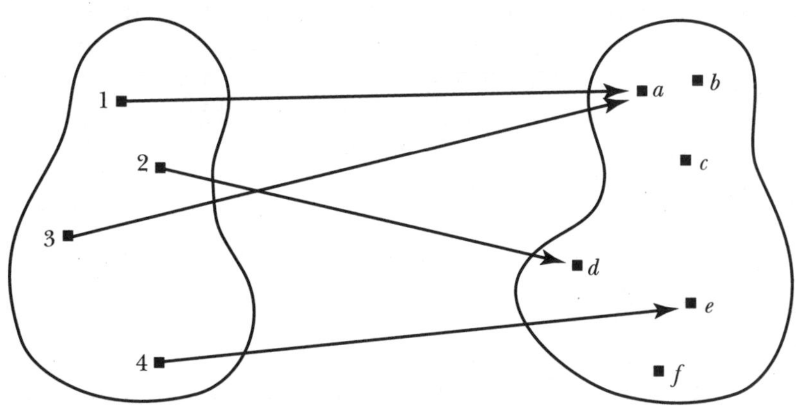

1. Assume f is the name of the function represented by the arrow diagram given above.

a. Find $f(2)$ and $f(4)$.

b. Find the range of f.

c. Represent f by a table.

2. Represent each of the following discrete functions by an arrow diagram.

a.

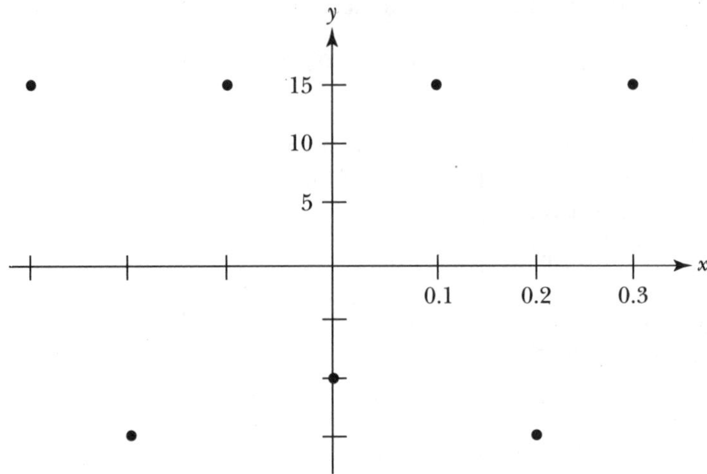

b. NumVowels := {["Anne",2], ["Elizabeth",4], ["Seth",1], ["Jamal",2], ["Tad",1]};

c. The function assigns each of the twelve months in a non-leap-year into the number of days in the month.

- In Unit 1 you used the vertical line test to determine if a graph represents a function. This approach does not apply, however, if the function is represented by something other than a graph. In general, to determine if a relationship defines a function you have to check whether each item in the domain corresponds to exactly one item in the range. None of the relationships given in the next exercise represents a function. Your job is to explain why.

HW2.3 Explain why each of the following relationships does not represent a function.

1. The relationship defined by the table:

x	"a"	"z"	"1"	"a"	"t"	"0"	"?"	"c"	"m"
y	6	−2	1	0	−4	1	8	6	6

2. The relationship defined by the set of ordered pairs:

F := {[2.3, 5.8], [6.7, 9.1], [5.8, 2.3], [7.1, 5,8], [6.7, 2.3]}

3. The relationship defined by the graph:

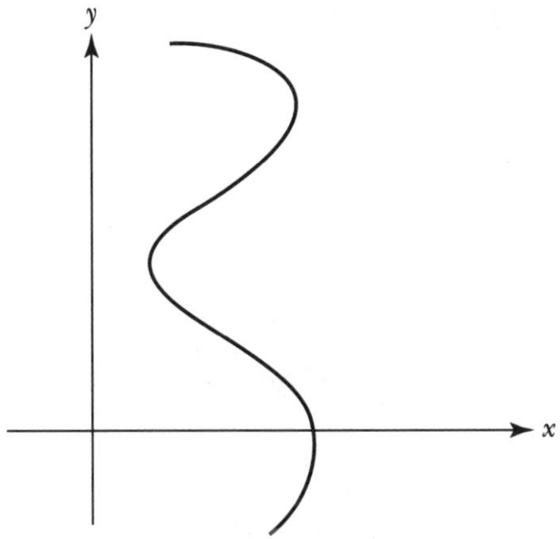

4. The relationship defined by the arrow diagram:

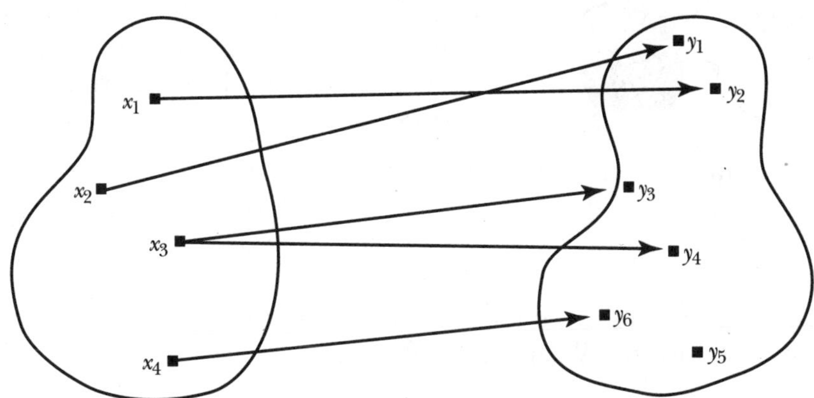

5. The relationship given by the piecewise definition:

$$h(r) = \begin{cases} 6, & \text{if } -3 \leq x \leq 0 \\ x^2, & \text{if } 0 \leq x \leq 4 \\ 2x, & \text{if } 4 \leq x \leq 6 \end{cases}$$

2

6. The relationship defined by the scatter plot:

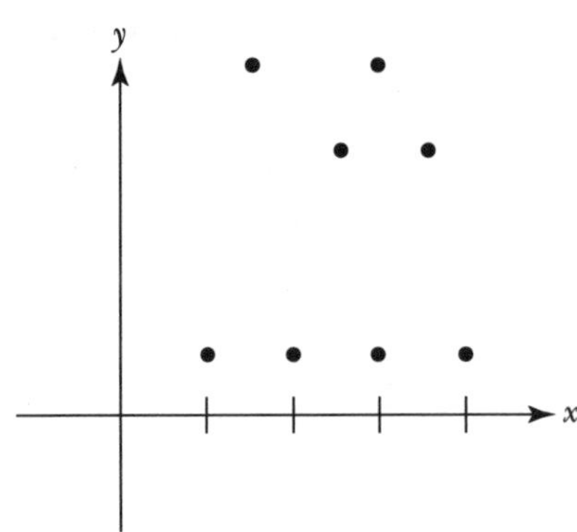

7. The relationship defined by the verbal description:

   Ladd loves chocolate covered almonds, but he knows that he has to exercise self-control or else he will eat too many. So on Sunday he allows himself to have 20, and then 2 less on each successive day of the week; unless he does not have a date for Saturday evening, in which case he lets himself have 25 on Saturday. Assume the domain is the 7 days of the week, Sunday through Saturday.

8. The graph given by the equation:

$$x^2 + y^2 = 1$$

- Develop an intuitive understanding of continuity and classify points of discontinuity in HW2.4.

HW2.4 The word "continuous" means going on without interruption. An intuitive way to test if a function is *continuous* over its domain is to trace the graph representing the function with a pencil. If you can do so without lifting the pencil, it is a continuous function. On the other hand, if you must lift the pencil at a point in the domain, then the function is *discontinuous* at that point since you cannot continue "without interruption."

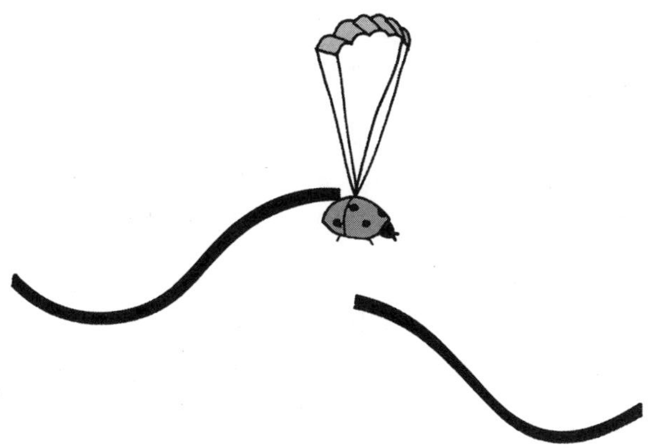

A function is said to have

- a *removable discontinuity* or a *hole* if the function can be redefined so that it is continuous at that point;
- a *jump discontinuity* if the graph has a vertical gap at that point (see the diagram given above);
- a *blow-up discontinuity* if the function has a vertical asymptote at that point—that is, if the values of the function explode as you approach the point from either the left and/or the right.

In order to use the do-I-or-don't-I-have-to-lift-the-pencil approach you need to know what the graph of the function looks like, which is not always the case. Moreover, to determine whether a function is continuous throughout its domain, you need to know what the entire graph looks like, which may not be possible if, for example, its domain is the set of all real numbers. To actually prove that a function is continuous you need another approach. You will use the tools of calculus to do this. In the meantime, use the intuitive approach.

1. For each of the following functions:

i. Graph the function. Identify the points in the domain (if any) where the function fails to be continuous by looking at the graph.

ii. Classify each point where the function is discontinuous as a removable, jump, or blow-up discontinuity.

a. $f(x) = \begin{cases} -x+2, & \text{if } x \le 1 \\ x^2, & \text{if } x > 1 \end{cases}$ **b.** $s(t) = \begin{cases} t^2, & \text{if } t < 0 \\ 4, & \text{if } t = 0 \\ t, & \text{if } t > 0 \end{cases}$

c. $h(t) = \begin{cases} t^2, & \text{if } t \le 0 \\ -t, & \text{if } 0 < t \le 4 \\ -5, & \text{if } t > 4 \end{cases}$ d. $f(x) = \begin{cases} 1/x, & \text{if } x > 0 \\ 0, & \text{if } x = 0 \end{cases}$

2. The *greatest integer function* assigns to each real number the largest integer less than or equal to that number. Denote the greatest integer in x by $[x]$.

a. Find [3.14], [7], and [−2.6].

b. Sketch $f(x) = [x]$ for $-10 \le x \le 10$.

c. Give the range of f.

d. List all the points where f is discontinuous. What types of discontinuity occur?

e. The greatest integer function is predefined in ISETL by the **floor** function. Explain why naming the function "floor" makes sense.

3. Similarly, $c(x) = [x]$ returns the *smallest integer* greater than or equal to x.

a. Find [3.14], [7], and [−2.6].

b. Sketch the graph of c for $-10 \le x \le 10$.

c. Give the range of c.

d. List all the points where c is discontinuous. What types of discontinuity occur?

e. c is predefined in ISETL by **ceil**, which is short for ceiling function. Explain why naming the function "ceiling" makes sense.

4. The cost of parking your car at Denver International Airport (DIA) is free for 0 to 70 minutes, $4 for 71 minutes to 2 hours, $2 for each additional half-hour or fraction thereof, to a maximum of $10 for the first 24 hours.

a. How much does it cost to park for 3 hr. 45 min.? 6 hr. 20 min.?

b. Sketch a graph of the cost function for the first day—that is, for $0 \le t \le 24$ hours.

c. Give the range of the function.

d. List all the points where the function is discontinuous. Classify each point of discontinuity.

e. Represent the cost function for parking at DIA with a piecewise-defined function.

- Explain what a function is in HW2.5.

HW2.5 What is a function?

Write an essay explaining what a function is. Your essay should include descriptions of (1) independent versus dependent variables, (2) domain and range, (3) discrete versus continuous functions, and (4) at least three different ways to represent a function and how to determine if each representation is a function.

SECTION 2

Combining Functions

Just as you can combine numbers by adding, subtracting, multiplying, and dividing to get a new number, you can combine functions to form a new function. This is particularly useful if you want to model a complex situation which can be expressed as a combination of separate operations. For example, suppose you are the chief financial officer (CFO) of a corporation consisting of 10 companies. Assume that for each of the companies, you have a function that models the earnings of that company over time. As CFO, however, you need to analyze the earnings of the entire corporation, not just the earnings of the individual companies. One way to do this is to create a model for your corporate earnings by adding the functions modeling the earnings for each company. The questions then are: What does it mean to "add" functions? How can you use the representations of the underlying functions to represent the new function? How is the domain of the new function related to the domains of the underlying functions?

Not only can you create new functions by adding two existing functions together but also by finding their difference, product, quotient, and composition, and by multiplying a function by a scalar, or a fixed real number. In order to do this, you need to know how to evaluate the new function at a point. For instance, if f and g are two functions, then you can add them to form the new function $f + g$, but how do you find the value of $(f + g)(x)$? The natural way to do this is to evaluate f at x and g at x and add the results. This leads to the following definition of the sum of two functions.

Sum: $$(f + g)(x) = f(x) + g(x)$$

Similarly, you can form new functions $f - g$, $f\,g$, f/g, and $c\,f$ where c is a scalar, using the following definitions:

Difference: $$(f - g)(x) = f(x) - g(x)$$

Product: $$(f\,g)(x) = f(x) \cdot g(x)$$

Quotient: $\left(\dfrac{f}{g}\right)(x) = \dfrac{f(x)}{g(x)}$

Scalar multiple: $(c\,f)(x) = c \cdot f(x)$ where c is a scalar.

The domains of these new functions are determined by the domains of f and g.

Before exploring how to represent a combination and determine its domain, evaluate some combinations at some given inputs and explain why the new function is not defined at some other inputs.

Task 2-5: Evaluating Combinations of Functions

Note: Although this is a pencil and paper exercise, you may want to use ISETL to check your results.

1. Consider the functions

$$f(x) = x^2 - 1, \quad \text{where } x \text{ is any real number}$$

$$g(x) = x, \quad \text{where } x \geq 0$$

a. Evaluate the following combinations of f and g at the given points. Show how you arrive at your results. For example, to evaluate $(f+g)(4)$, you might write

$$\begin{aligned} (f+g)(4) &= f(4) + g(4) && \text{use definition of sum} \\ &= 3 + 2 && \text{evaluate } f \text{ and } g \text{ at } x = 4 \\ (f+g)(4) &= 5 && \text{simplify} \end{aligned}$$

(1) $(f - 4g)(25)$

(2) $\left(\dfrac{f}{g}\right)(9)$

(3) $(fg)(6)$

(4) $(6.5g)(100)$

b. Explain why it is not possible to evaluate each of the following combinations.

(1) $(gf)(-1)$

(2) $\left(\frac{g}{f}\right)(1)$

2. Consider the functions

x	−2	−1	0	1	2	3	4
f(x)	0	5	3	1	−1	4	−1

x	−2	0	2	4	6	8	10
g(x)	3	−4	2	0	6	3	1

a. Evaluate the following combinations. Use the definitions of the various combinations to show how you arrive at your results.

(1) $(f - g)(-2)$

(2) $(2f + 10g)(0)$

(3) $\left(\frac{g}{6f}\right)(4)$

(4) $(gf)(-2)$

b. Explain why it is not possible to evaluate the following combinations.

(1) $(f + g)(8)$

(2) $\dfrac{f}{g}(1)$

(3) $\dfrac{f}{g}(4)$

3. Consider the functions

$$f(x) = \begin{cases} 2x + 6, & \text{if } x \geq 0 \\ x - 8, & \text{if } x < 0 \end{cases}$$

$$g(x) = x^3, \quad \text{where } x \text{ is any real number}$$

a. Evaluate the following combinations. Use the definitions to show how you arrive at your results.

(1) $(f - g)(-2)$

(2) $(2f + 10g)(0)$

(3) $\dfrac{g}{6f}(4)$

(4) $(gf)(-2)$

b. Explain why it is impossible to evaluate $(f/9g)(0)$.

(2) Domain:

b. $2f$

(1) Representation:

$$(2f)(x) =$$

(2) Domain:

c. $\frac{f}{g}$

(1) Representation:

$$\left(\frac{f}{g}\right)(x) =$$

(2) Domain:

2. Consider the functions:

x	−2	−1	0	1	2	3	4
$f(x)$	0	5	3	1	−1	4	−1

x	−2	0	2	4	6	8	10
$g(x)$	3	−4	2	0	6	3	1

Represent each of the following combinations by a table. Give the domain of the combination.

a. $f + g$

(1) Representation:

x	
$(f + g)(x)$	

(2) Domain:

b. $-4\,fg$

(1) Representation:

x	
$(-4f + g)(x)$	

(2) Domain:

c. $\dfrac{f}{g}$

(1) Representation:

x	
$\left(\frac{f}{g}\right)(x)$	

(2) Domain:

2

ii. Using the points from part i, sketch the graphs representing $2f$ and $\frac{1}{2}f$ on the same axes as f. Label the two graphs.

(a)

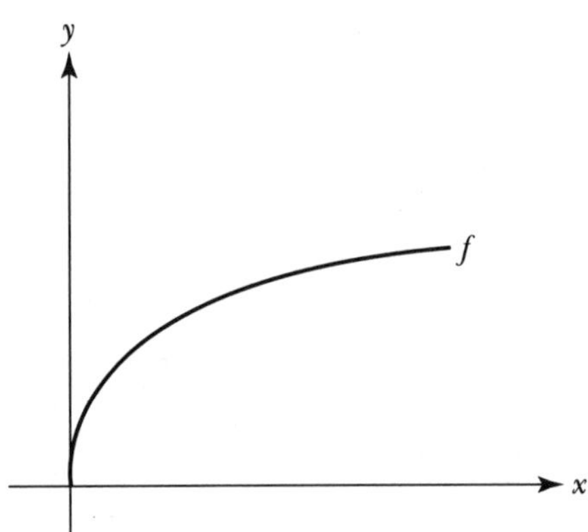

(b)

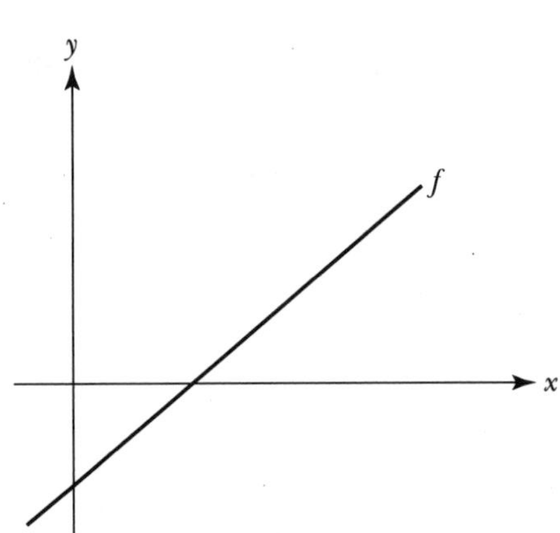

b. Represent the sum of two functions—which are represented by graphs—by a graph.

(1) Recall that by definition $(f + g)(x) = f(x) + g(x)$.

Suppose $P(a, f(a))$ is the point on the graph of f, and $Q(a, g(a))$ is the point on the graph of g corresponding to $x = a$. Approximate the value of $f(a) + g(a)$. Use this approximation to locate the point on the graph of $f + g$ corresponding to $x = a$. Label the point R.

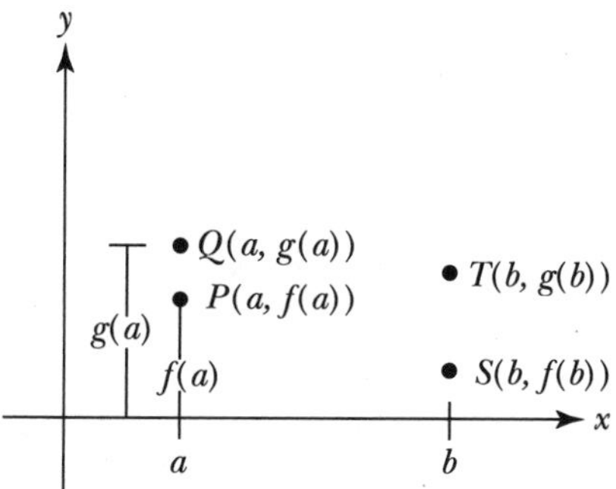

Similarly, suppose S and T are points on the graphs of f and g respectively corresponding to $x = b$. Locate the point on the graph of $f + g$ corresponding to $x = b$. Label the point U.

(2) Explain how in general you can use the graphs of f and g to determine the graph of $f + g$.

(3) Try it. Suppose the functions f and g are represented by graphs. In each case,

i. Mark several x values on the horizontal axis. For each value, indicate the value of $f(x)$, $g(x)$, and $(f + g)(x)$.

ii. Using the points from part i, sketch the graph representing $f + g$ on the same axes as f and g.

(a)

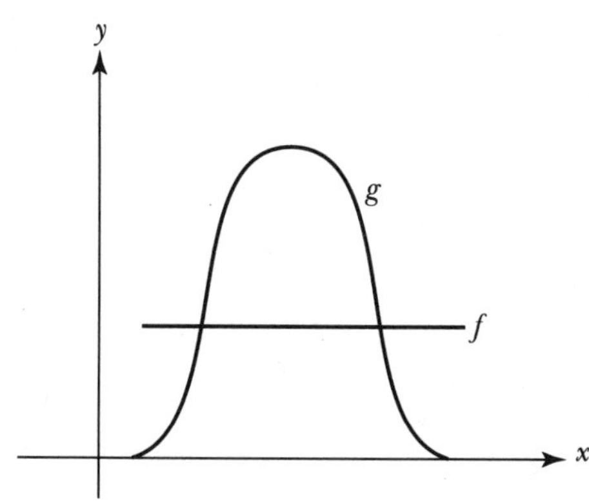

2

(3) $g \circ f(-0.5)$

(4) $g \circ f(4)$

(5) $g \circ f(0)$

b. Consider the functions

x	−2	−1	0	1	2	3	4
$f(x)$	0	5	3	1	−1	4	−1

x	−2	0	2	4	6	8	10
$g(x)$	3	−4	2	0	6	3	1

If possible, evaluate $g \circ f$ at the given input, showing how you arrived at your result. If $g \circ f(x)$ does not exist for the given value of x, explain why.

(1) $g \circ f(-2)$

(2) $g \circ f(4)$

(3) $g \circ f(0)$

(4) $g \circ f(8)$

(5) $g \circ f(3)$

c. Consider the functions

$$f(x) = \begin{cases} 2x + 6, & \text{if } x \geq 0 \\ x - 8, & \text{if } x < 0 \end{cases}$$

$$g(x) = x^3, \quad \text{where } x \text{ is any real number}$$

If possible, evaluate $g \circ f$ at the given input. Show how you arrived at your result. If $g \circ f(x)$ does not exist, explain why.

(1) $g \circ f(-2)$

(2) $g \circ f(4)$

(3) $g \circ f(0)$

(4) $g \circ f(8)$

2. Represent a composition and find its domain.

In the following exercises, represent the composition using the same representation as the underlying functions. For instance, if both f and g are represented by tables, represent $g \circ f$ by a table.

Determine the domain of the composition. Note that the domain of $g \circ f$ is contained in the domain of f, since f is evaluated first. The question then is: What values in the domain of f are not in the domain of the com-

d. $h(x) = |x - 1.5|$, where x is any real number.

e. $h(x) = \sqrt{2x - 1}$, where $x \geq 0.5$.

Unit 2 Homework After Section 2

- Complete the tasks in Section Two in the Activity Guide. Be prepared to discuss them in class.
- Evaluate some combinations and find some representations in HW2.6.

HW2.6

1. Consider the functions

$$f(x) = \frac{1}{x}, \quad \text{where} \quad x \neq 0 \qquad \text{and} \qquad g(x) = \begin{cases} x + 1, & \text{if } x < 3 \\ -x, & \text{if } x > 3 \end{cases}$$

a. Evaluate each of the following combinations. If it is not possible, explain why.

(1) $(f + 3g)(1)$

(2) $\left(\frac{f}{g}\right)\left(\frac{1}{2}\right)$

(3) $(g \circ f)\left(\frac{1}{3}\right)$

b. Represent each of the following combinations by a piecewise-defined function.

(1) $\frac{2g}{f}$

(2) $f \circ g$

2. Consider the functions

r	−1	0	1	2	3	4	5
$h(r)$	3	4	5	6	0	1	3

and

r	0	1	2	3	4	5	6
$m(r)$	0	−4	1	4	2	−1	3

a. Evaluate each of the following combinations. If it is not possible, explain why.

(1) $(m - 5h)(4)$
(2) $(hm)(3)$
(3) $\left(\frac{2h}{m}\right)(0)$
(4) $(m \circ h)(2)$

b. Represent each of the following combinations by a table.

(1) $m + h$
(2) $h \circ m$

3. Represent the following functions as the composition of two functions.

a. $a(x) = \dfrac{1}{x^2 + 3x + 2}$
b. $b(x) = (x^4 + 4x^2 + 2x)^3$
c. $c(x) = \cos(x^2 + 7)$
d. $d(x) = 9 - (x + 9)^9$

4. Let $f(x) = x^3$, $g(x) = \sqrt{x}$, $h(x) = x - 4$, and $j(x) = 2x$. Express each of the following functions as a composition of three of these four functions:

a. $k(x) = 2(x - 4)^3$
b. $k(x) = \sqrt{(x - 4)^3}$
c. $k(x) = (2x - 8)^3$
d. $k(x) = 2\sqrt{x - 4}$

5. Consider the following diagram:

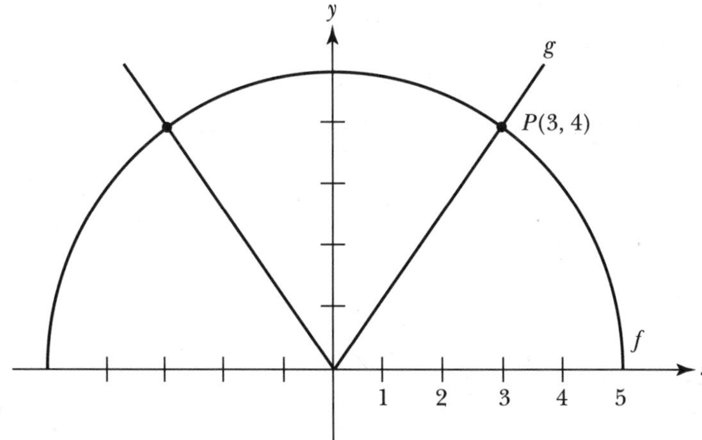

a. Find:

(1) $(f + g)(0)$
(2) $(f + g)(3)$
(3) $(f + g)(5)$
(4) $(f + g)(-3)$

2

Task 2-8: Sketching Reflections

Before beginning this task, look carefully at the figure given in the discussion at the beginning of this section. Observe that each point P on the graph of the given function corresponds to a point Q on the graph of the reflection, where Q can be obtained by moving in a direction perpendicular to the line of reflection and where the distance from P to the line of reflection equals the distance from Q to the line of reflection.

In the following exercises:

i. Find the reflection of a few points on the given graph.

ii. Sketch the graph of the reflection. Label the reflection.

1.

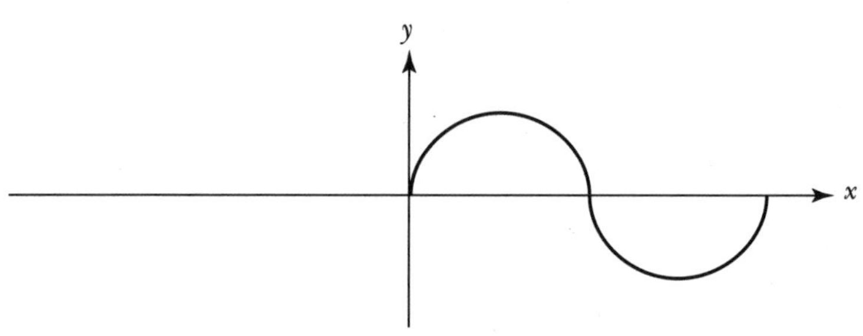

a. Reflect the given function through the x-axis. Label.

b. Reflect the given function through the y-axis. Label.

2.

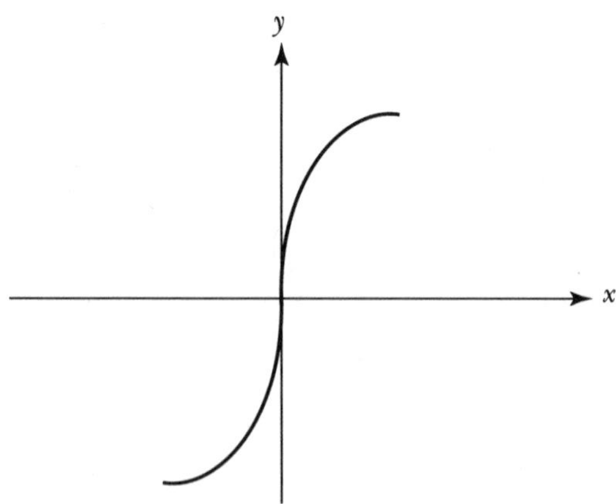

a. Reflect the given function through the x-axis. Label.

b. Reflect the given function through the y-axis. Label.

3. Reflect the following function through the x-axis.

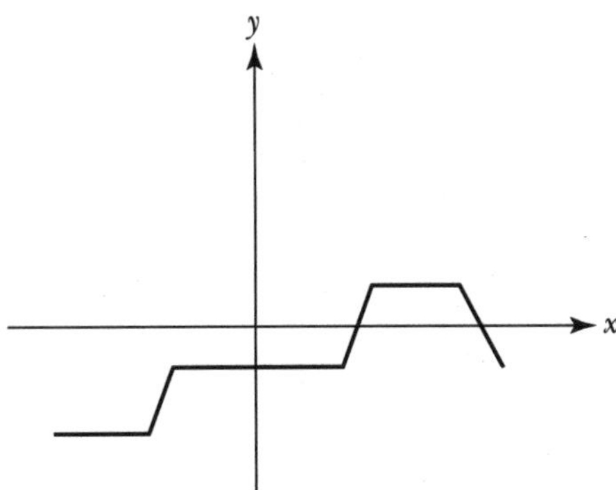

4. Reflect the function given in part 3 through the y-axis.

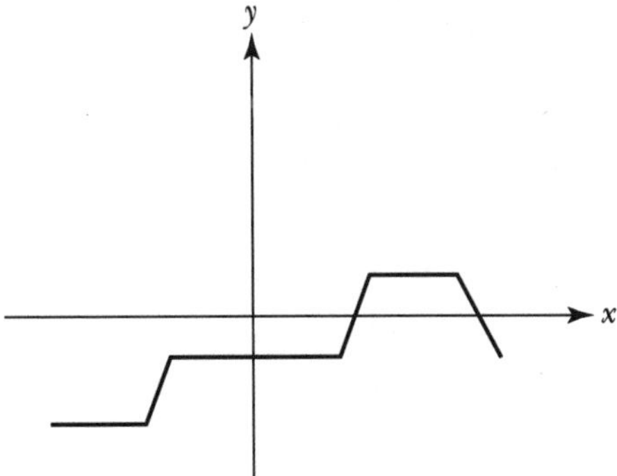

5.

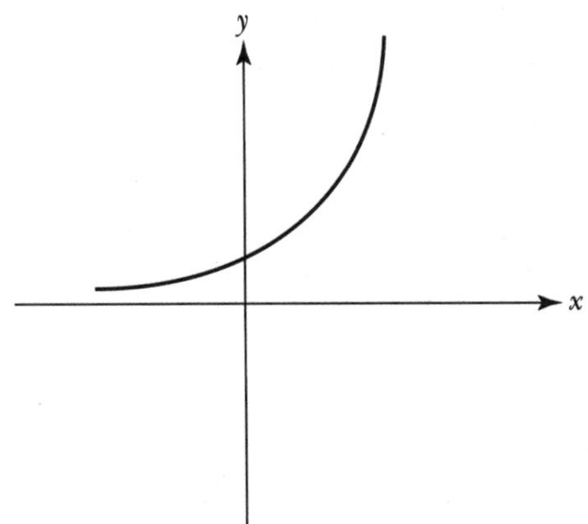

a. Reflect the given function through the x-axis. Label.

b. Reflect the given function through the y-axis. Label.

(a) Carefully plot the points P_1 to P_6 on the axes given below. Label the axes and label each point.

(b) Suppose the points Q_1 to Q_6 are the reflections of P_1 to P_6 through the y-axis. Find the coordinates of the reflections. Record your information in the table.

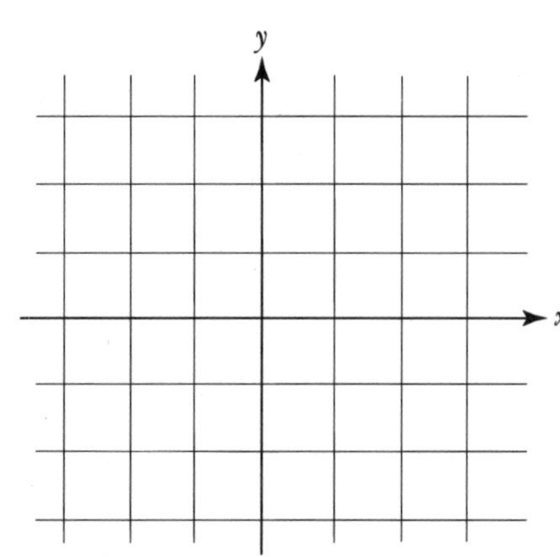

Point on Given Function	Reflection Through y-axis
$P_1(-2,2)$	$Q_1(\ ,\ )$
$P_2(-1,0)$	$Q_2(\ ,\ )$
$P_3(0,3)$	$Q_3(\ ,\ )$
$P_4(1,-3)$	$Q_4(\ ,\ )$
$P_5(2,1)$	$Q_5(\ ,\ )$
$P_6(3,-2)$	$Q_6(\ ,\ )$

(2) Generalize your observations.

(a) Suppose the point Q is the reflection of the point $P(x_0,y_0)$ through the y-axis. Find the coordinates of Q in terms of x_0 and y_0; that is, fill in

$$P(x_0,y_0) \leftrightarrow Q(\quad,\quad)$$

(b) Describe the relationship between the coordinates of a point and the coordinates of its reflection through the y-axis.

b. Find an expression for the reflection of a function through the vertical axis and examine how this expression is related to the expression representing the given function.

Based on your work in part a, you know you can find the coordinates of the reflection of any point through the vertical axis by negating its first coordinate—that is, by replacing the value of x with $-x$. You can use the same approach to find the expression for the reflection; replace every x in the given expression with $-x$, and simplify (if possible). Try it.

(1) Consider $y = 2x + 1$. Show that $y = 2(-x) + 1$, or $y = -2x + 1$, is the reflection of $y = 2x + 1$ through the y-axis by graphing the two functions on the same axes.

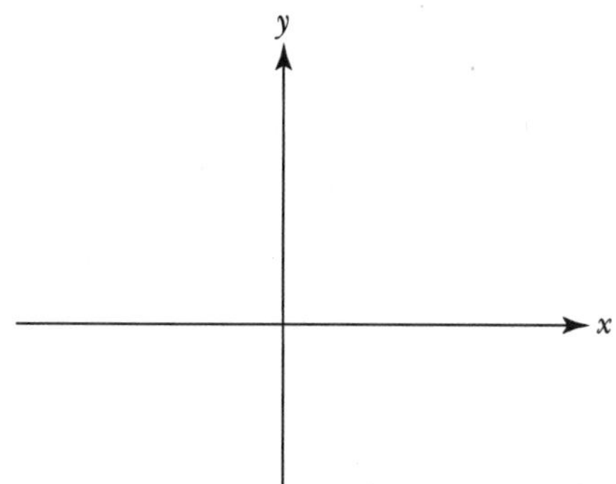

2

(2) Consider $y = \sqrt{x}$, where $x \geq 0$.

(a) Find an expression for the reflection of the given function through the y-axis.

Caution: Be careful with the domain. You also have to replace x with −x here.

(b) Draw rough sketches of the given function and its reflection through the y-axis on the axes given below. Label the graphs.

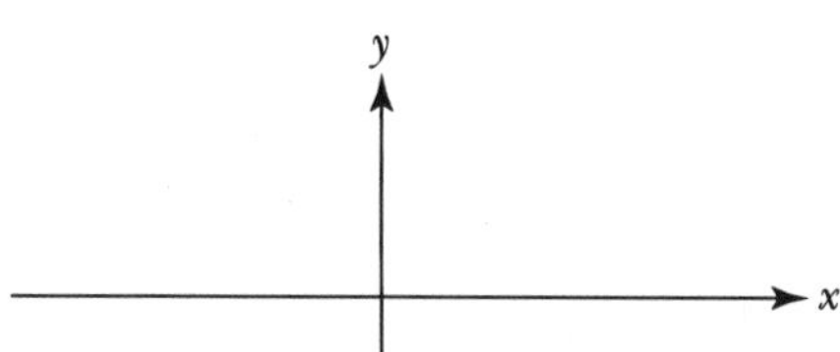

c. Remember that the graph of a function satisfies the Vertical Line Test. Is it possible for the reflection of a function through the y-axis to fail this test? In other words, is the reflection of a function through the y-axis always a function? Justify your response.

d. In calculus, functions are usually given a name; for example, $f(x) = 2x + 1$ or $f(x) = \sqrt{x}$. In this case, replacing x with $-x$ is equiv-

(2) Consider $y = \sqrt{x}$, where $x \geq 0$.

(a) Find an expression for the reflection of the given function through the x-axis. What is the domain of the reflection in this case?

(b) Draw rough sketches of the given function and its reflection through the x-axis on the axes given below. Label the graphs.

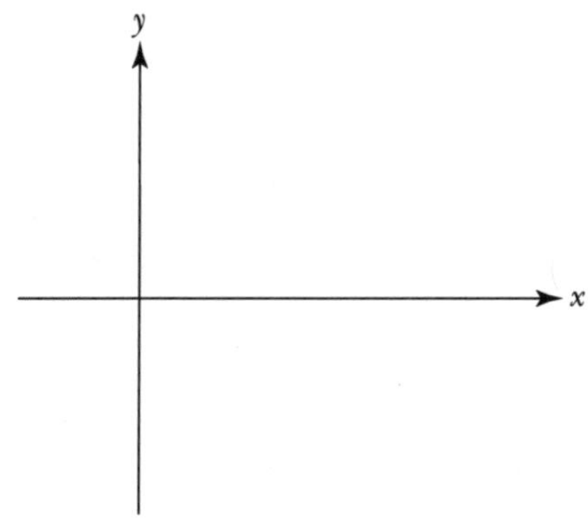

c. The domain of the reflection of a function through the horizontal axis is always the same as the domain of the given function. Explain why this is the case.

d. Is the reflection of a function through the x-axis always a function? Justify your response.

e. If the function has a name, such as f, then $y = f(x)$. In this case, replacing y with $-y$ is equivalent to finding $-f(x)$. Represent the reflection of the following functions through the x-axis by an expression.

(1) $f(x) = 3x^2 - 6x + 9$, where x is any real number.

(2) $f(x) = x^2$, where $x \geq 0$.

(3) $f(x) = \dfrac{1}{x+1}$, where $x \neq -1$.

3. Finally, consider the reflection of a function through the line $y = x$.

a. Reflect some points.

(1) Determine the relationship between the coordinates of a given point and the coordinates of the reflection of the point through the line $y = x$. Choose your own points. Find their reflections. Record your data on the graph and in the table given below. Figure out what happens in this case.

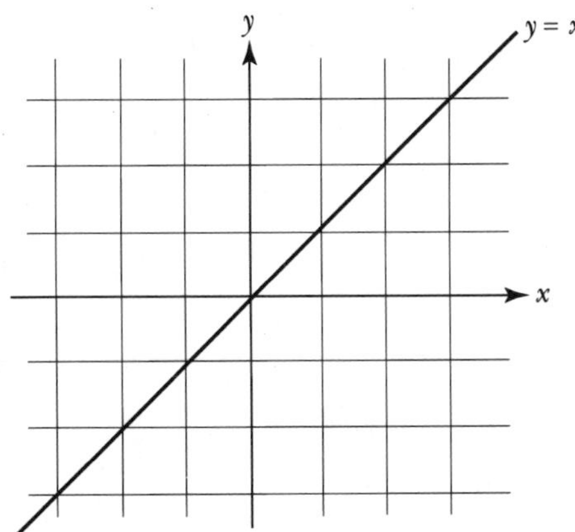

Point on Given Function	Reflection Through $y = x$

(2) Generalize your observations.

(a) Suppose Q is the reflection of the point $P(x_0,y_0)$ through the line $y = x$. Find the coordinates of Q in terms of x_0 and y_0.

$$P(x_0,y_0) \leftrightarrow Q(\quad , \quad)$$

Before continuing on, let's summarize what you discovered in the last task.

When you *reflect a function through the y-axis,* each point $P(x_0,y_0)$ on the graph of a given function corresponds to the point $Q(-x_0,y_0)$ on the graph of its reflection, as shown in the diagram below. To represent the reflection by an expression, you replace each x in the expression for the given function with $-x$, or if the function has a name, you find $f(-x)$.

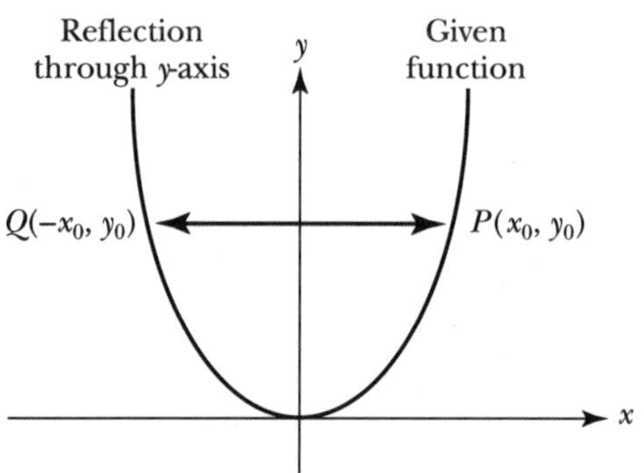

When you *reflect a function through the x-axis,* each point $P(x_0,y_0)$ on the graph of a given function corresponds to the point $Q(x_0 - y_0)$ on the graph of its reflection. To represent the reflection by an expression, you replace each y in the expression for the given function with $-y$ and solve for y, or if the function has a name, you find $-f(x)$.

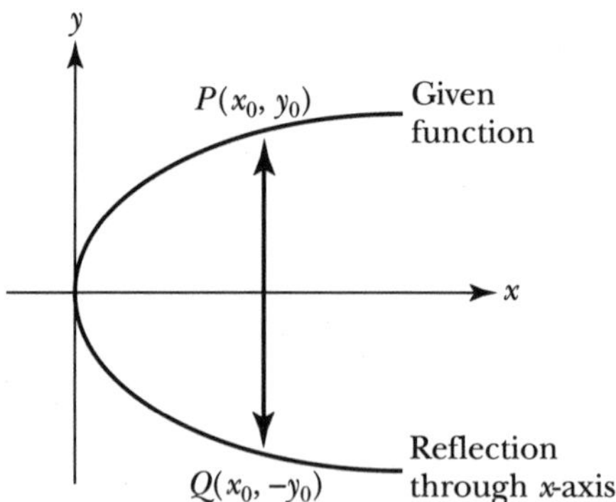

Finally, when you *reflect a function through the line* $y = x$, each point $P(x_0,y_0)$ on the graph of a given function corresponds to the point $Q(x_0,y_0)$ on the graph of its reflection. To represent the reflection by an expression, you interchange x and y in the expression for the given function and solve for y.

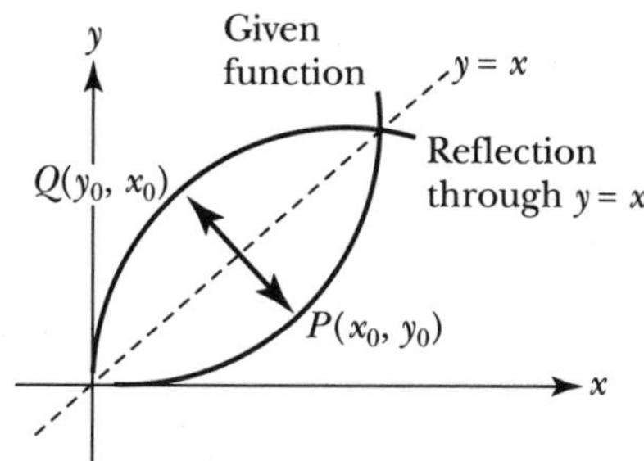

The reflection through the line $y = x$ is probably the most useful type of reflection since it enables you to reverse, or undo, the process of a function. There is one problem, however: The reflection of a function through the line $y = x$ is not necessarily itself a function. For example, the reflection of $y = x^3$ is a function, whereas the of reflection of $y = x^2$ is not. What is it about these two functions that causes this to happen?

In the next task, you will discover conditions guaranteeing that the reflection of a function through $y = x$ is itself a function. This new function is said to be the *inverse* of the given function. Since the inverse of a function reverses the process of the original function, composing a function with its inverse gets you back to where you started.

2

Task 2-10: Investigating Inverse Functions

1. Examine some situations where the reflection through $y = x$ is not a function. Think about what causes this to occur.

Recall that a relationship defines a function if corresponding to each input (x value) there exists exactly one output (y value). A graph represents a function if it passes the Vertical Line Test—that is, if every vertical line passes through at most one point on the graph.

Consider the graph of the function $y = x^2$.

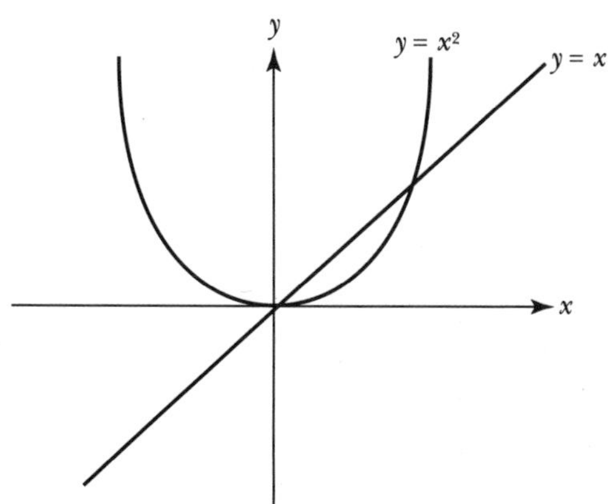

2. If the reflection of a given function through the line $y = x$ is also a function, the reflection is called the *inverse* of the given function. In this case, the given function is said to be a *one-to-one* function, since not only does each x value correspond to exactly one y value, but each y value corresponds to exactly one x value.

Determine if each of the following functions has an inverse—that is, if the given function is one-to-one. If it is not, show why not. If it is, represent the inverse using the same form as the given function.

a.

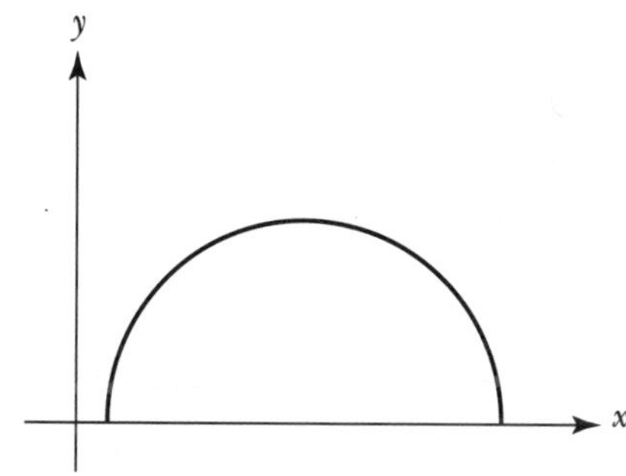

b. $y = -2x$, where $0 \leq x \leq 3$.

c.

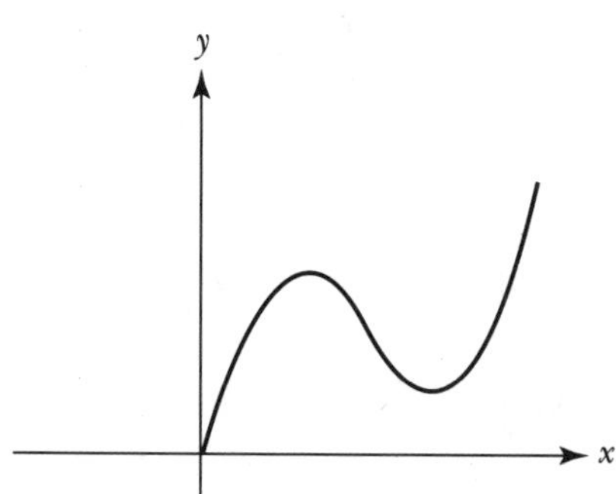

d.

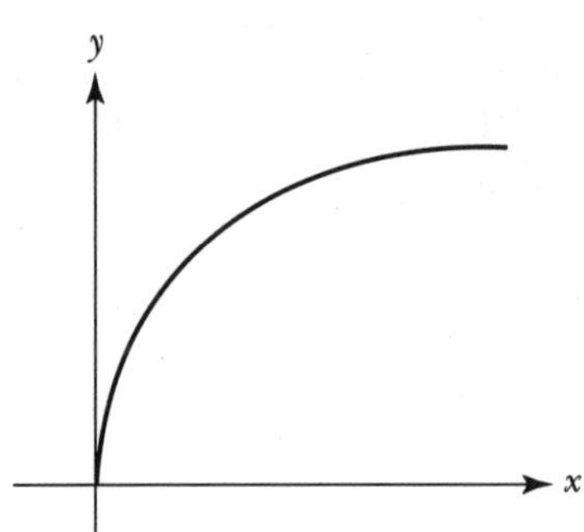

3. The inverse of a function reverses, or undoes, the process of the given function. As a result, the composition of a function and its inverse returns the original input value. Moreover, if g is the inverse of f, then f is the inverse of g. Consequently,

$$f \circ g(x) = f(g(x)) = x, \text{ for all } x \text{ in the domain of } g$$

and

$$g \circ f(x) = g(f(x)) = x, \text{ for all } x \text{ in the domain of } f$$

a. Show that $f(x) = 3x - 1$ and $g(x) = \frac{1}{3}x + \frac{1}{3}$ are inverses of each other by showing that $f \circ g(x) = x$ and $g \circ f(x) = x$, for any real number.

b. The formula to convert a temperature t from Celsius to Fahrenheit is given by $F(t) = \frac{9}{5}t + 32$.

(1) Find the formula for converting a temperature t from Fahrenheit to Celsius.

$C(t) =$

(2) Show that F and C are inverses of each other.

4. If a function f has an inverse, its inverse is denoted by f^{-1}. Consequently, $f^{-1} \circ f(x) = x$ and $f \circ f^{-1}(x) = x$.

Caution: Do not confuse inverse functions with reciprocal functions. Just as $4^{-1} = \frac{1}{4}$, $(f(x))^{-1}$, denotes the reciprocal of $f(x)$; that is,

$$(f(x))^{-1} = \frac{1}{f(x)}$$

In contrast, $f^{-1}(x)$ denotes the inverse of f evaluated at x. In general, these are different functions. Be careful not to mix up their notations. To determine which is which, observe where the superscript "-1" appears.

Suppose f has an inverse, where $f(0) = -6$ and $f(-1) = 2$. Find the value of:

a. $f^{-1}(-6)$

b. $(f(-6))^{-1}$

c. $f^{-1}(f(0))$

d. $f \circ f^{-1}(2)$

e. $f^{-1} \circ f(10)$

f. $-(f(0))^{-1}$

Unit 2 Homework After Section 3

- Complete the tasks in Section Three in the Activity Guide. Be prepared to discuss them in class.
- Find the reflection of some functions in the HW2.9 and HW2.10.

HW2.9 Consider the four functions:

i. $f(x) = x^2 + 1$, where $x \geq 0$.

ii. $f(x) = x$, where x is any real number.

iii. $f(x) = \quad x - 1$, where $x \geq 0$.

iv. $f(x) = -1$ where $x \geq 0$.

1. Recall that $y = f(x)$. Reflect each of the functions given above in i–iv through the y-axis.

   a. Find a formula for the reflection.

   b. Sketch the function and its reflection on a single pair of axes. Label the two graphs.

   c. Give the domain of the reflection.

2. Reflect each of the functions given above in i–iv through the x-axis.

   a. Find a formula for the reflection.

   b. Sketch the function and its reflection on a single pair of axes. Label the two graphs.

HW2.10 Consider the functions:

i. $f(x) = -2x + 4$, where x is any real number.

ii. $f(x) = 1$, where $x \leq 0$.

For each of the functions given above in i–ii:

1. Sketch the function and its reflection through the line $y = x$ on a single pair of axes. Label the two graphs.

2. Determine whether or not f has an inverse. If it does not, explain why. If f^{-1} does exist:

   a. Represent f^{-1} by an expression and find the domain of f^{-1}.

b. Show that $f \circ f^{-1}(x) = x$, for all x in the domain of f^{-1}.

c. Show that $f^{-1} \circ f(x) = x$, for all x in the domain of f.

• Answer some questions concerning reflections in HW2.11.

HW2.11 Think in general about reflecting functions, as you respond to the following questions.

1. Describe the shape of a function whose reflection through the horizontal axis is the function itself—that is, when you reflect the function through the horizontal axis, you get the same function. Support your description by sketching the graph of a few functions where this is the case.

2. Describe the shape of a function whose reflection through the vertical axis is the function itself. Support your description by sketching the graph of a few functions where this is the case.

3. Describe the shape of a function whose reflection through the horizontal axis is the same as the reflection of the function through the vertical axis. Support your description by sketching the graph of a few functions where this is the case.

4. If the domain of a given function is the set of all numbers greater than or equal to zero, find:

a. the domain of the reflection of the function through the y-axis.

b. the domain of the reflection of the function through the x-axis.

5. Suppose f is a constantly increasing function.

a. Explain why f must have an inverse.

b. Is f^{-1} an increasing or a decreasing function? Justify your answer and support it with two examples.

c. If f is concave up, is f^{-1} concave up or concave down?

6. Describe the arrow diagram for a function that has an inverse.

• Invert some functions which are not represented by expressions in HW2.12.

HW2.12 To reflect a function represented by an expression through the line $y = x$, you interchange the values of x and y. This approach can be generalized to functions represented in other forms by interchanging the associated input and output values.

1. Invert the functions given in parts a–e. Represent the reflection using the same representation as the given function. For instance, in part a, f is represented by a table; represent its reflection by a table.

a.

x	"a"	"c"	"f"	"i"	"m"	"u"	"z"
f(x)	5.4	−0.2	2.6	−1.5	2.4	0.2	8.5

b. $f = \{[1,2], [3,4], [5,-9], [7,-7], [9,-5]\}$

c. The function represented by the arrow diagram:

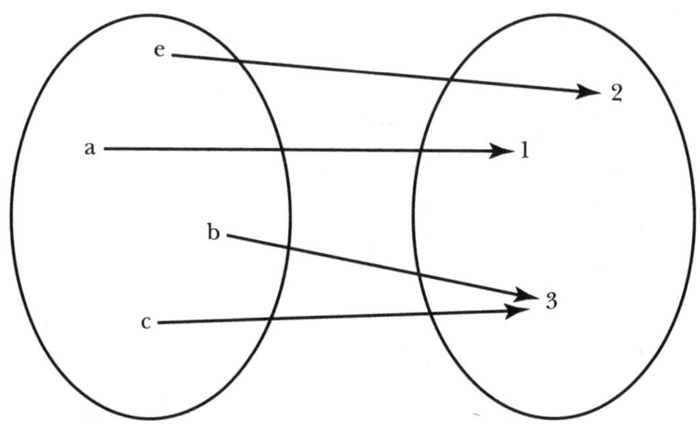

d. The function represented by the scatter plot:

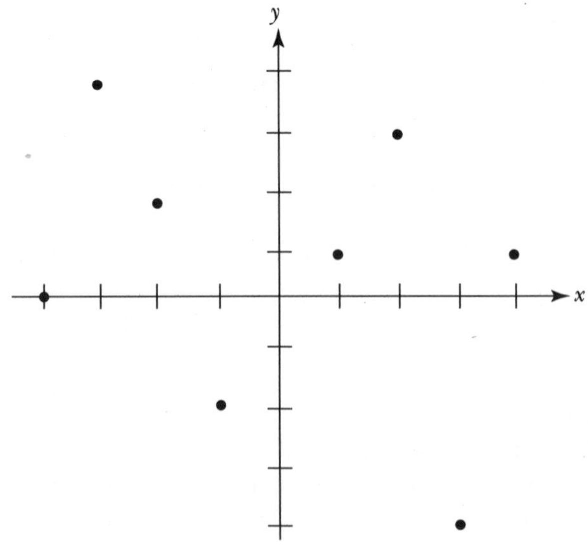

2. Which reflections in part 1 are functions?

- The Vertical Line Test gives a way of testing whether or not a given graph represents a function. The Horizontal Line Test, on the other hand, gives a way of testing whether or not a given graph represents a function that is one-to-one—that is, if each y value corresponds to exactly one x value. Explain why the Horizontal Line Test makes sense in HW2.13.

HW2.13 The Horizontal Line Test.

1. According to the Horizontal Line Test:

> A graph represents a function that is one-to-one if and only if every horizontal line in the xy-plane intersects the graph in at most one point.

Explain why this makes sense. Support your explanation by sketching two appropriate diagrams (one which represents a one-to-one function and one which does not).

2. For each of the following graphs, use the Horizontal Line Test to determine whether the function is one-to-one and, consequently, whether it has an inverse.

a.

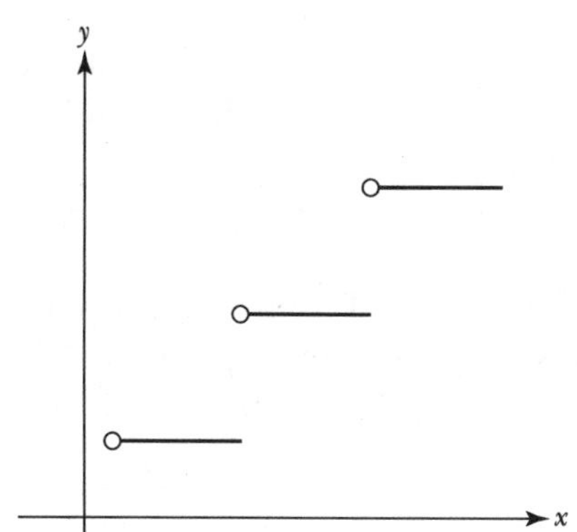

b.

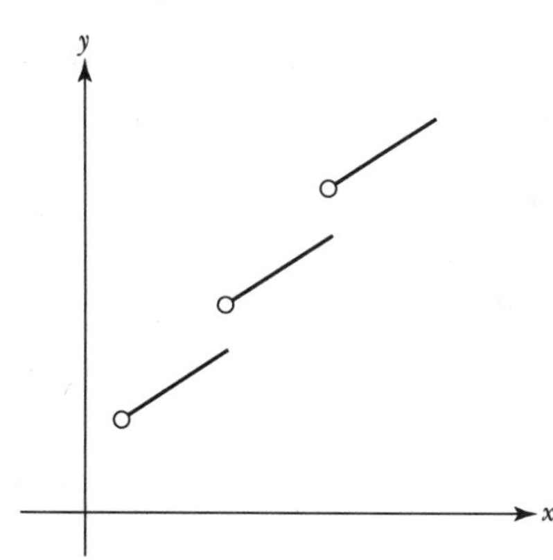

c.

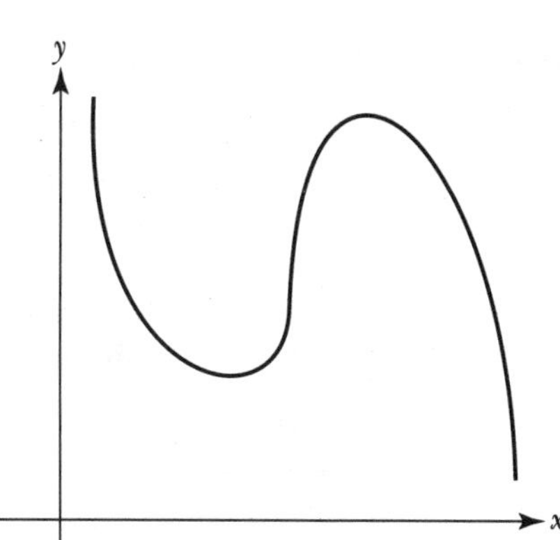

d.

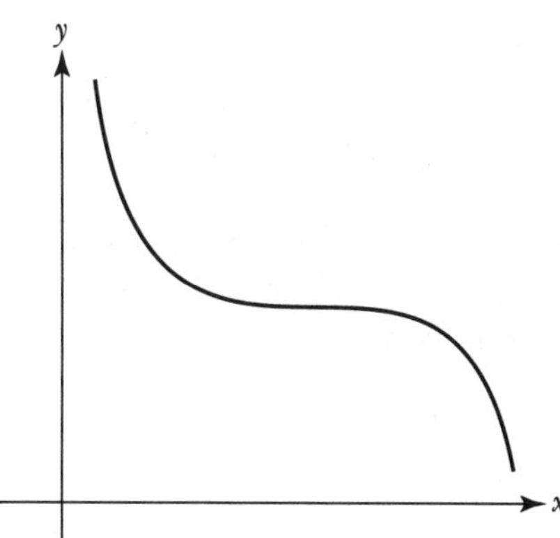

2

3. If $g(-4) = 1$ and $g(0) = 1$, use the Horizontal Line test to explain why g does not have an inverse.

4. Remember your beer-selling uncle (see HW2.1, part 3)? The number of liters of beer he sells at a Saturday game is a function of the temperature, where

$$NumLiters(t) = 354 + 8(t - 62), \text{ for } 62°\text{F} \leq t \leq 103°\text{F}.$$

   a. Explain why *NumLiters* has an inverse.

   b. Find an expression for the inverse of *NumLiters*. Call it *Temp*.

   c. Find *Temp*(498). Describe what this value represents.

   d. Find the domain and range of *Temp*.

   e. Prove that your expression for *Temp* is correct by showing that
   (1) $Temp \circ NumLiters(t) = t$, for $62°\text{F} \leq t \leq 103°\text{F}$.
   (2) $NumLiters \circ Temp(t)$, for all t in the domain of *Temp*.

- Determine whether or not some functions are one-to-one in HW2.14.

HW2.14 A function is one-to-one if each output value corresponds to exactly one input value. Determine whether or not each of the following verbal descriptions represents a function that is one-to-one. Justify your response.

1. You manage a shop that sells 200 different small items. You never keep more than 100 of a particular item in stock. Your inventory control function maps each item in your shop to the number of that item that you currently have in stock.

2. The domain of the function is the set of all people in Pennsylvania, and the range is a subset of the positive integers. The function assigns each person to his or her age.

3. The domain of the function is the set of all people in the world. The range is a subset of all pairs of men and women. The function maps a person to his or her natural-parents.

4. The domain of the function is the set of all people. The range is the set of all right thumb prints. The function maps a person to his or her right thumb print.

5. The domain of the function is the set of all current students at Kirch College. The range is the set of all student mailbox numbers. The function assigns a student to his or her mailbox.

6. The function represents your movement in front of a motion detector over a 10-second time interval. Starting $\frac{1}{2}$ meter from the detector, you slowly increase your distance from the detector for 5 seconds; you stand still for 3 seconds; and then move rapidly towards the detector during the remaining 2 seconds.

7. The function represents your movement in front of a motion detector over a 7-second time interval. Starting 4 meters from the detector, you decrease your distance from the detector. You start walking very slowly, and then walk faster and faster.

- Write your journal entry for this unit. Before you begin to write, review the material in the unit. Think about how it all fits together. Try to identify what, if anything, is still causing you trouble.

HW2.15 Write your journal entry for Unit 2.

1. Reflect on what you have learned in this unit. Describe in your own words the concepts that you studied and what you learned about them. How do they fit together? What concepts were easy? Hard? What were the main (important) ideas? Give some examples of the main ideas.

2. Reflect on the learning environment for the course. Describe the aspects of this unit and the learning environment that helped you understand the concepts you studied. What activities did you like? Dislike?

Name ______________________________ Section __________ Date ______

Unit 3:

FUNCTION CLASSES

Mathematics is no longer just an entry-level prerequisite for engineering, the physical sciences, and statistics. Its principles and techniques, along with computers, have become a part of almost all areas of work; and its logic is used in thinking about almost everything. This is a big change from the days when a number of professions were virtually math-free. Today, many occupations that do not require college-level calculus or statistical skills at the outset demand them later on for anyone aiming toward promotion into management or work in more interesting technical areas.

Sheila Tobias, *Succeed with Math*, p. 4,
The College Board, 1987.

OBJECTIVES

1. Examine various function classes, including:

   - polynomial
   - rational
   - trigonometric
   - exponential
   - logarithmic

2. Investigate the shape and location of the associated graphs.
3. Inspect the limiting behavior of functions near holes and asymptotes, and for large values in the domain.

4. Compare growth rates of functions with similar shapes.
5. Fit a curve to a set of data and use the model to analyze the data and make predictions.

OVERVIEW

The goal of the first two units was to help you develop a general understanding of the meaning of a function.

- You considered a variety of functions represented by everything from squiggly curves and randomly generated tables to ones represented by specific expressions.
- You described a function's domain, range, and process, and determined if it was one-to-one.
- You identified the intervals where a function was increasing/decreasing and concave up/down, and located the function's local maxima and minima.
- You examined the behavior of the tangent line as it traversed a smooth curve.

In this unit, you will examine some specific types of functions or *function classes*, including polynomial, rational, trigonometric, exponential, and logarithmic functions. Some of these, such as $f(x) = mx + b$ or $f(\theta) = \sin(\theta)$ will be familiar to you, whereas others, such as $f(t) = \ln(t)$, may be new.

You will analyze the graphs of the functions and develop mental images associated with their shape and location. As a result, by the time you finish this unit, statements such as "the population is growing exponentially" or "the length of the day over the course of a year is sinusoidal" will conjure up particular images in your mind. In addition, you will observe how varying the coefficients of functions in a particular class affects the graphs of the functions. For example, you will compare the graph of $\sin(\theta)$ to the graphs of functions such as $2\sin(\theta)$, $\sin(2\theta)$, $\sin(\theta) + 2$, and $\sin(\theta - 2)$. You will investigate the behavior of a rational function near a hole or a vertical asymptote, and for large values of x. You will examine the relationship between functions, such as $\log_a x$ and a^x or $\cos(\theta)$ and $\sec(\theta)$. You will compare the growth of functions whose graphs have similar shapes, such as the growth of $x^2 + 1$ versus the growth of 2^x. As you examine these issues, you will convert back and forth between verbal, symbolic, and graphical representations of a function and use functions to model real-life situations.

Much of the work you will do will be done using pencil and paper. A CAS (*computer algebra system*), however, such as Maple, Derive, Theorist, or Mathematica, can do almost all the things you can do by hand, and more. In particular, not only can a CAS perform basic numeric operations, it can also perform symbolic manipulations, such as factoring a polynomial or expanding a collection of terms, and it can graph functions in two and three dimensions. In this unit you will become familiar with your CAS and use it to analyze functions.

Using a CAS will enable you to tackle more complex problems, and it will help prevent you from getting stuck part way through a problem, either because you can't figure out how to do the algebra or because you have made an algebraic mistake somewhere along the way. More importantly, using a CAS will enable you to concentrate on how to solve a problem, instead of concentrating on details, such as plotting numerous points to graph a function or performing a lengthy list of messy algebraic calculations to reach a solution. Although a CAS can do much of the work for you, you still have to tell it exactly what to do. Consequently, to use your CAS effectively, you have to understand what needs to be done and you must be able to describe how to do it in a precise, logical manner. With a CAS the challenge of solving a problem is in the understanding, not the doing.

SECTION 1

Polynomial and Rational Functions

3

You can tell that a function belongs to a particular class by looking at its format. A polynomial, for instance, is a combination of integer powers of x. In general, a *polynomial of degree n* has the form

$$f(x) = a_n x^n + a_{n-1} x^{n-1} + \cdots + a_1 x + a_0$$

where n is a nonnegative integer, the a_i's are real numbers, and $a_n \neq 0$. Consequently, all of the following are polynomials, since each one has the specified form. (As you look through the list, determine the degree of each polynomial.)

$$\begin{aligned} f(x) &= 3x^3 - 4x^2 + 9x + 15 \\ g(r) &= r^{100} \\ k(x) &= 2.5x^{10} - 3x^5 + 1.1 \\ h(t) &= 26.78 \end{aligned}$$

Some polynomials can be categorized based on the value of their degree. For instance,

Degree	Category	Example
0	Constant	$f(x) = -6.78$
1	Linear	$m(t) = 2t - 8.5$
2	Quadratic	$d(s) = 6s^2 - 4.5s + 99$
3	Cubic	$p(x) = x^3 - 2x + 1.4$

Polynomials are old friends. When you took algebra, you sketched their graphs by plotting points. You factored and simplified their expressions. You found their x- and y-intercepts. Given a verbal description of a polynomial, you represented it symbolically and vice versa. For instance, given information about a line, you found its corresponding expression, and given the equation for a line, you determined whether it rose or fell, described its steepness, and indicated where it intersected the vertical axis. In the next task, you will examine some of these ideas again. Later in this section, you will analyze the behavior of rational functions near a hole or a vertical asymptote, and you will learn how to use your computer to help you do this.

Task 3-1: Examining Polynomial Functions

1. The domain of any polynomial is the set of all real numbers. Explain why this is the case by considering its general form.

2. Examine the place where the graph of a polynomial crosses the vertical axis.

   Note: The y-intercept of a function is the value where the graph of the function crosses the vertical axis. If y_0 is the y-intercept of a function f, then $f(0) = y_0$, and the graph of f passes through the point $P(0,y_0)$.

a. For each of the following polynomials find the value of its y-intercept. Mark its location on the vertical axis.

(1) $f(x) = 3x^3 - 4x^2 + 9x + 15$

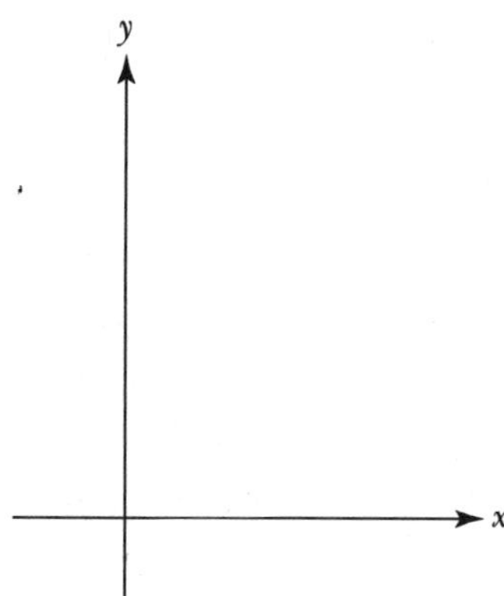

(2) $g(x) = 2.5x^{10} - 3x^5 + 1.1$

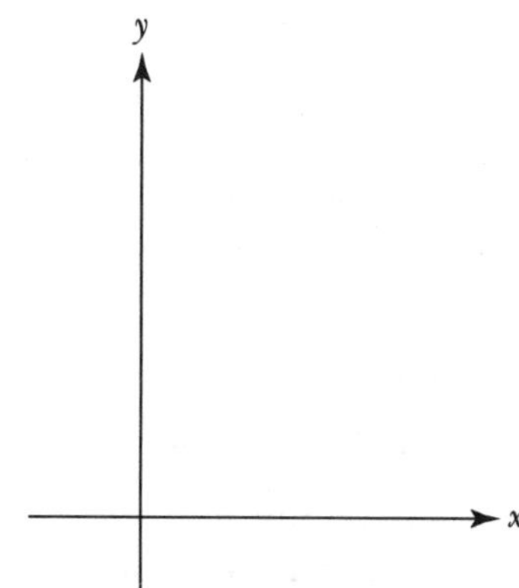

(3) $h(t) = -6.78$

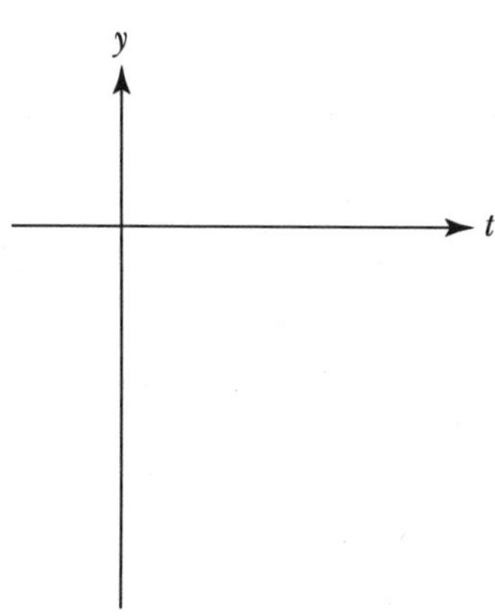

b. A polynomial has one, and only one, y-intercept. Explain why this is always the case. (First explain why a polynomial must have a y-intercept, and then explain why it cannot have more than one.)

3. Examine the place(s) where the graph of a polynomial crosses the horizontal axis.

Note: A polynomial crosses the vertical axis exactly one time. On the other hand, it might intersect—touch or cross—the horizontal axis once, twice, 10 times, or maybe not at all. Mathematically speaking, an x-intercept of a function is a value where the graph of the function intersects the horizontal axis. If x_0 is an x-intercept of a function f, then $f(x_0) = 0$, and the graph of f passes through the point $P(x_0,0)$. An x-intercept is also referred to as a zero or a root of the function.

a. If x_0 is a root of a polynomial function, then $(x - x_0)$ is a factor of the function. Consequently, one way to find the roots of the function is to:

i. Factor the polynomial expression.

ii. Set each of the factors equal to 0 and solve for x.

(1) Find all the roots of the following polynomials by factoring. Show your work. Use this information to mark the x-intercepts on the given horizontal axis.

(a) $g(x) = x^2 - 5x - 6$

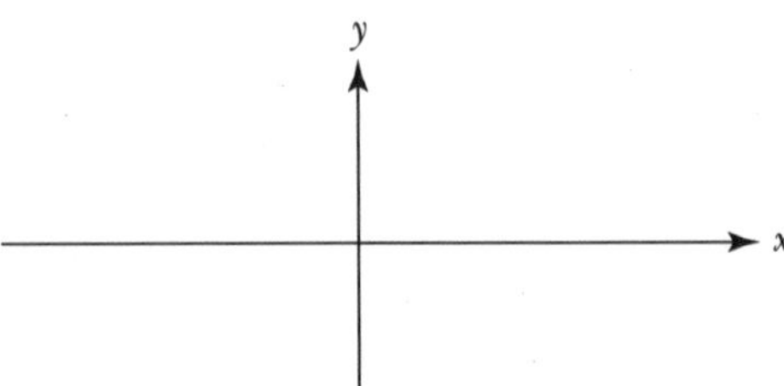

(b) $p(x) = 3x^3 - 3x$

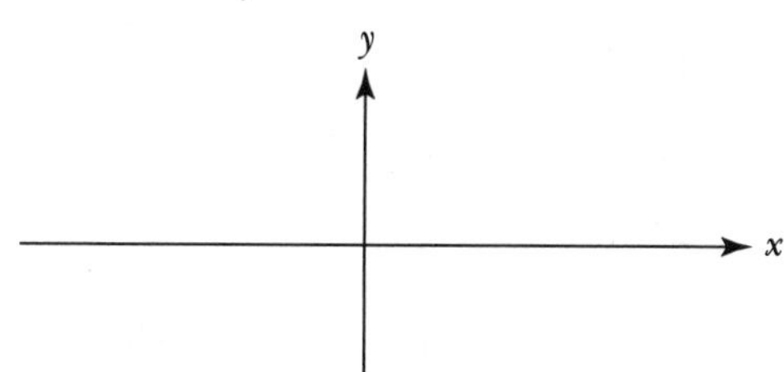

(c) $h(x) = (x - 0.5)^5$

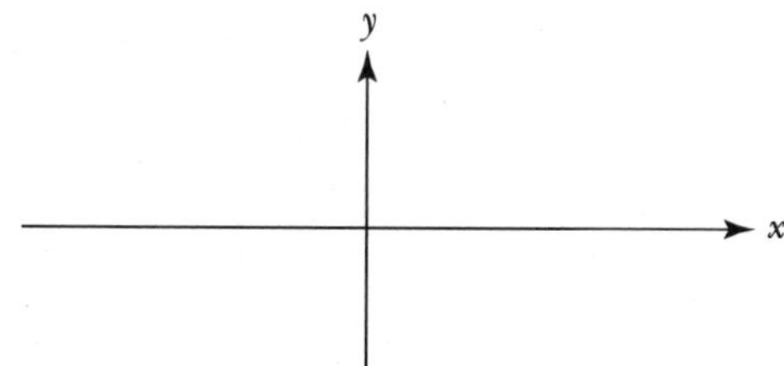

(d) $m(x) = 26.78$

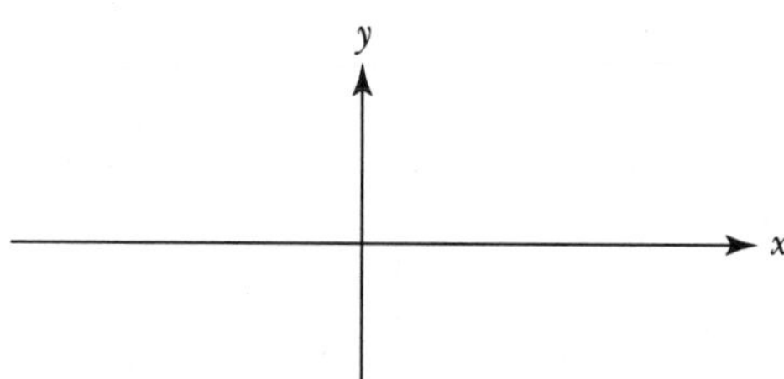

(e) $y = \frac{5}{6}x - 10$

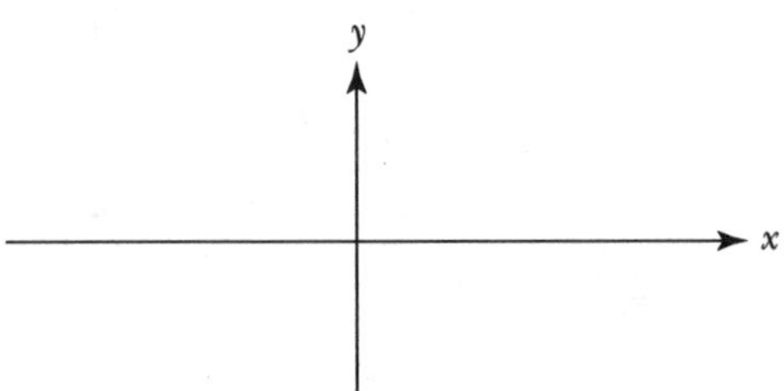

(f) $q(x) = x^2 - 5x + 6$

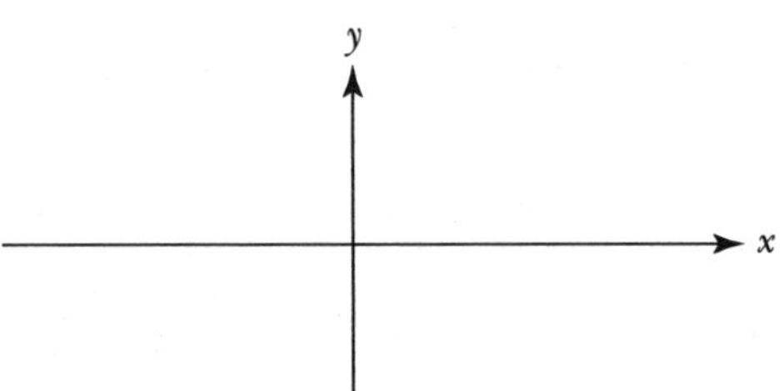

(2) Explain why the method of factoring method works; that is, explain why factoring a polynomial and setting its factors to 0 gives all the roots of the polynomial.

b. In the case of a second degree polynomial—that is, a polynomial which has the form

$$f(x) = ax^2 + bx + c, \text{ where } a \neq 0$$

—you can use the *quadratic formula* to find its roots where

$$x = \frac{-b \pm \sqrt{b^2 - 4ac}}{2a}$$

(1) Show that the quadratic formula yields the same results as factoring.

Consider $f(x) = x^2 + 2x - 3$.

(a) Use the quadratic formula to find the roots of f.

(b) Use factoring to find the roots of f. Your results should be the same as in part (a). If not, check your work.

(2) If the roots are not integers, it is almost impossible to factor a quadratic function by hand. The quadratic formula, on the other hand, always works.

i. Find the roots of the following polynomials by using the quadratic formula. Show your work.

ii. Use the roots to express each function as a product of its factors.

(a) $q(x) = -x^2 + 5x - 4$

(b) $f(x) = 3x^2 + 5x - 1$

(c) $r(x) = -4x^2 + 4x - 1$

(d) $s(x) = -2x^2 + 0.5x - 1$

(3) The graph of a quadratic function, $f(x) = ax^2 + bx + c$, where $a \neq 0$, is always shaped like a "U" and is called a *parabola.* As a result, there are three possible scenarios: its graph does not intersect the horizontal axis, in which case the function has no real roots; its graph touches the horizontal axis at one point, in which case the function has one real root; or its graph intersects the axis at two distinct points, in which case it has two real roots.

The part of the quadratic formula which is under the square root sign, $b^2 - 4ac$, is called the *discriminant.* The sign of the discriminant indicates whether the quadratic function has no, one, or two real roots. For each of the cases given below:

i. Give the sign of the discriminant: positive $(+)$, negative $(-)$, or zero (0).

ii. Explain why this sign implies that a quadratic function has the specified number of roots.

iii. Sketch two quadratic functions that have the given number of roots.

(a) Case 1: no real roots.

(b) Case 2: one real root.

(c) Case 3: two real roots.

4. Examine the impact of the coefficients on the shape and location of the graphs of linear and quadratic functions.

   a. A first degree polynomial is called a *linear function* and has the form $f(x) = mx + b$, where $m \neq 0$. Examine how the values of the coefficients m and b affect the graph of the function.

      (1) Describe the general shape of the graph of a first degree polynomial.

      (2) Describe the impact the sign of m has with respect to whether a linear function is increasing or decreasing. Support your claims with some sample sketches.

      (3) Describe what the size of the value of m (large positive versus small positive, more negative versus less negative) tells you about the steepness of the graph of the function. Support your description with some sample sketches.

      (4) Describe what the value of b tells you about the graph of the function. Support your description with a sketch.

(5) Describe the relationship between two linear functions that have the same value for their leading coefficient, m. Support your description with a sketch.

(6) Find an equation for the x-intercept of $f(x) = mx + b$, where $m \neq 0$, in terms of m and b.

(7) What happens if $m = 0$? Give the general form for the polynomial. Describe the shape and location of the graph. Support your description with some sample sketches.

3

b. Recall that the graph of a quadratic function, $f(x) = ax^2 + bx + c$, where $a \neq 0$, is always a parabola. As a result, the graph has exactly one turning point. The graph, however, can have a large variety shapes—for instance, it can be concave up or down, be skinny or fat, and have two, one, or no x-intercepts. Examine how the sign and the value of the coefficients a and c affect the shape and location of a parabola.

(1) On the axes given below, draw rough sketches of the graphs of $y = x^2$ and $y = -x^2$. Describe in general how the sign of the leading coefficient, a, affects the concavity of a quadratic function.

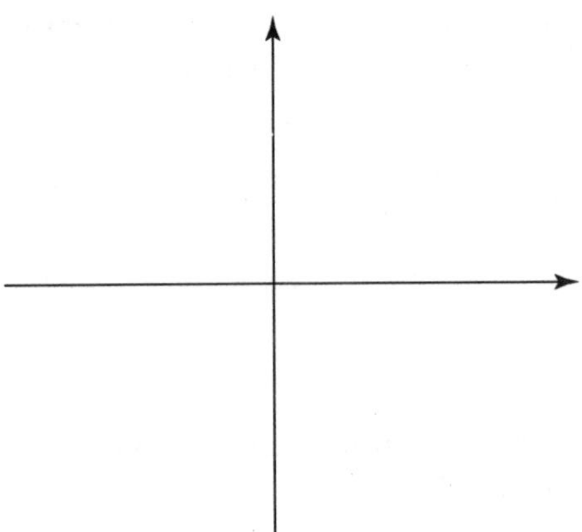

(2) On the axes given below, draw rough sketches of the graphs of $y = x^2 + 3$, $y = x^2 - 1$, and $y = x^2 + 1.5$. Describe how varying the value of c affects the location of the graph of $f(x) = x^2 + c$.

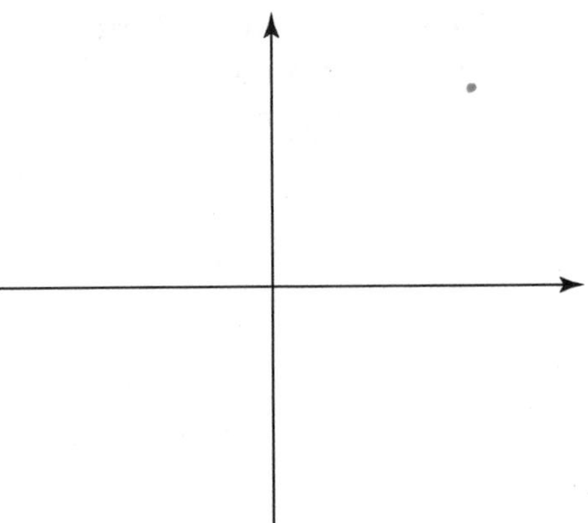

(3) On the axes given below, draw rough sketches of the graphs of $y = x^2$, $y = 2x^2$, and $y = x^2/2$. Describe how varying the value of a affects the shape of the graph of $f(x) = ax^2$.

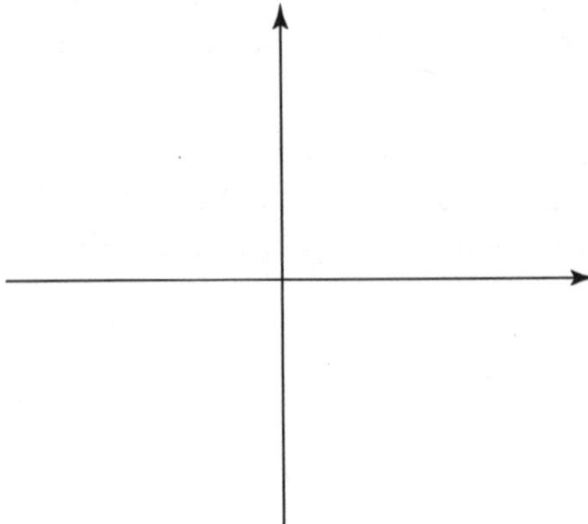

A *rational function* is the quotient of two polynomials. An important question when analyzing the behavior of a rational function is:

> What impact does a zero in the denominator of a rational function have on the shape of its graph?

For instance, neither of the following rational functions, r or s, is defined at $x = 3$, since you cannot divide by zero.

$$r(x) = \frac{x^2 - 2x - 3}{x - 3}$$

$$s(x) = \frac{x + 2}{x - 3}$$

The impact, however, of this zero in the denominator on the graphs of r and s is quite different; one graph has a *hole*, whereas the other has a *vertical asymptote.* In the next task, you will investigate why this is the case. You will also learn how to determine whether a zero in the denominator results in a hole or an asymptote by examining the expression for the rational function.

3

Task 3-2: Analyzing Rational Functions

1. Examine the behavior of a rational function near a hole.

Consider

$$r(x) = \frac{x^2 - 2x - 3}{x - 3}$$

for x near 3.

a. Give the domain of r.

b. Examine the limiting behavior of r as x approaches 3 from either side.

(1) Find some values of $r(x)$ as x approaches 3 from the left by using your calculator or computer to fill in the chart given on the following page.

x	r(x)
2.0	
2.2	
2.4	
2.6	
2.8	
2.9	
2.95	
3.00	om

Note: om indicates that r is not defined at $x = 3$.

(2) Find some values of $r(x)$ as x approaches 3 from the right as you fill in the chart given below.

x	r(x)
4.0	
3.8	
3.6	
3.4	
3.2	
3.1	
3.05	
3.00	om

(3) Based on the information in parts (1) and (2), what appears to be happening to the values of $r(x)$ as x gets close to 3 from either side?

c. Although $r(x)$ is not defined at $x = 3$, the values of $r(x)$ appear to get closer and closer to 4 as x gets closer and closer to 3. This tells you that the graph of r has a *hole* at $x = 3$. In fact, it even tells you where the hole is, namely at the point $P(3,4)$. Another way to show that r has a hole at $x = 3$ is to simplify the expression representing r and cancel the $x - 3$ term.

Simplify r and observe what happens.

Caution: The domain of the function is fixed at the outset. Simplifying the expression does not change the function's domain.

(1) Give the domain of r (again).

(2) Simplify the expression representing r by factoring the numerator and denominator and canceling like terms.

(3) Represent r using the simplified expression, along with the (original) domain.

$r(x) =$, where

(4) Use the simplified version to sketch the graph of r on the axes given below. Indicate the location of the hole with a small open circle.

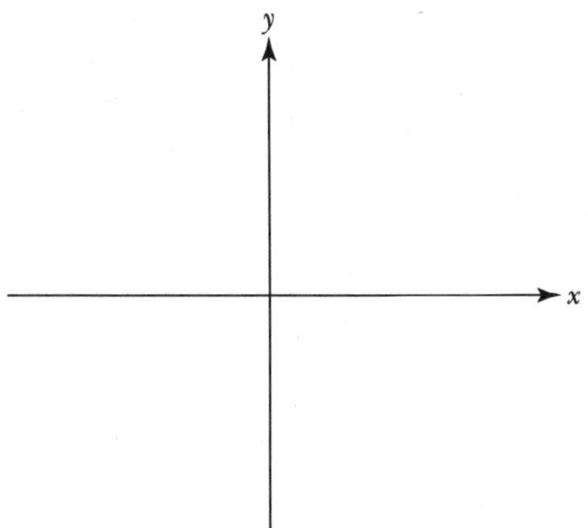

3

2. Examine the behavior of a rational function near a vertical asymptote.

In particular, analyze

$$s(x) = \frac{x+2}{x-3}$$

as x gets closer and closer to 3, and see how the behavior of s differs from the behavior of the function r (which you considered in part 1).

a. Give the domain of s.

b. In the case of r you were able to "get rid of" the problematic $x - 3$ term in the denominator. Explain why this approach does not work with s.

c. Examine the limiting behavior of s as x approaches 3 from either side.

(1) Find some values of $s(x)$ as x approaches 3 from the left and the right by filling in the following chart.

x	s(x)
2.96	
2.97	
2.98	
2.99	
3.00	om
3.01	
3.02	
3.03	
3.04	

(2) Describe what happens to the value of $s(x)$ as x approaches 3 from the left.

(3) Describe what happens to the value of $s(x)$ as x approaches 3 from the right.

d. s is not defined at $x = 3$. Moreover, the value of $s(x)$ "explodes" as x gets closer and closer to 3. This tells you that the graph of s has a *vertical asymptote* at $x = 3$. The values in the table in part c also indicate how the function s behaves as x gets closer and closer to the asymptote. In particular, when the asymptote is approached from the left, the function is decreasing since the value of $s(x)$ is approaching $-\infty$. On the other hand, when the asymptote is approached from the right, the function is increasing, since the value of $s(x)$ is approaching $+\infty$.

Sketch the graph of s for x near 3 on the axes given below.

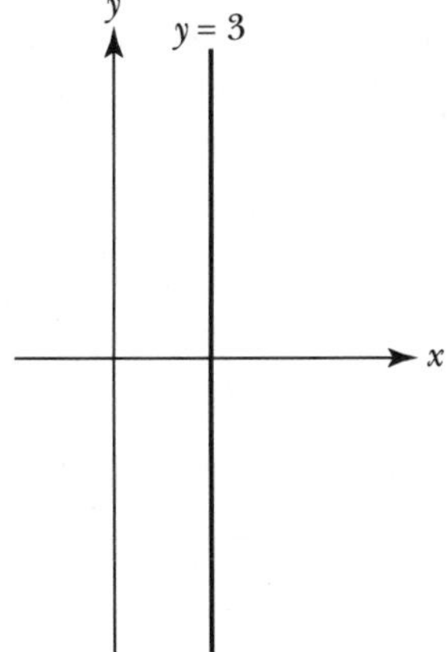

3. Not only can a rational function have vertical asymptotes, it can also have a *horizontal asymptote.* In this case, the values of the function get closer and closer to a fixed number as x gets farther and farther away from 0 in the positive direction and/or the negative direction.

It turns out that the function

$$s(x) = \frac{x+2}{x-3}$$

3

which you analyzed in part 2 has a horizontal asymptote. Investigate the behavior of s as x gets larger and larger in both a positive and negative sense.

a. Find some values of $s(x)$ as x approaches $+\infty$. Calculate the values of $s(x)$ to four decimal places.

x	10	100	1,000	10,000
s(x)				

b. Find some values of $s(x)$ as x approaches $-\infty$. Calculate the values of $s(x)$ to four decimal places.

x	−10	−100	−1,000	−10,000
s(x)				

c. Based on the two tables, what appears to be happening to the values of $s(x)$ as x gets larger and larger, in both the positive and negative directions?

4. Sketch the graph of the rational function s.

a. Reiterate your observations from the work you did in the previous two parts.

(1) s has a vertical asymptote at $x =$ _______.

(2) s has a horizontal asymptote at $y =$ _______.

b. Find the value of y where the graph of s intersects the y-axis—that is, find the y-intercept.

c. Find the values of x, if any, where the graph of s intersects the x-axis—that is, find the roots of s or the x-intercepts.

d. Sketch the graph of s, using the information from parts a, b, and c. Use dashed lines to indicate the locations of the vertical and horizontal asymptotes.

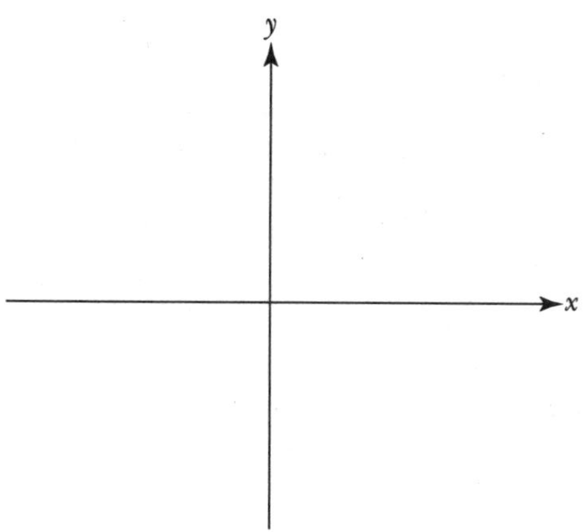

5. Summarize your observations.

Assume f is a rational function, where the numerator and denominator of f have been factored. Explain how you can use the factors in the numerator and the denominator to analyze the behavior of the function.

a. Describe a general method for finding the domain of f by considering the factors in the denominator of f.

b. Describe a general method for finding the holes of f by comparing the factors in the numerator and denominator of f.

c. Describe a general method for finding the vertical asymptotes of f by comparing the factors in the numerator and denominator of f.

3

d. Describe a general method for finding the zeros or x-intercepts of f by using the factors in the numerator of f.

e. Describe a general method for finding the y-intercept of f.

Analyzing the behavior of polynomial and rational functions involves doing a lot of symbolic manipulations and calculations. In the last two tasks, you did all of this by hand. Fortunately most of these types of things can be done using a computer. Using the computer will save you from getting bogged down in all the nitty-gritty details. It will force you, however, to think at a higher level, since in order to use the computer, you must be able to articulate a procedure for solving a problem. For instance, if you want to find the location of all the holes and vertical asymptotes of a given rational function what might you do? First, you could find the candidates for the holes and vertical asymptotes by finding all the values where the denominator of the given function is 0. One way to do this is to factor the denominator and find all its roots. Next you need to decide if a particular value results in a hole or an asymptote. Continuing to think generally, you know that if the term (which corresponds to the value under consideration) cancels with a like term in the numerator, it signifies a hole; if it does not cancel, then it defines a vertical asymptote. How can you determine whether or not a term cancels? One approach is to factor the numerator and look for like terms.

To use a computer to do all this, you do not need to know how to find the actual factors of the numerator and the denominator. However, you do need to know that factoring is an appropriate thing to do, so you can instruct the machine to do it for you. You then need to be able to interpret the results. In addition, you need to be able to judge if the results that the computer gives you are reasonable. For instance, if you ask the computer to factor a quadratic function and it returns three factors, when it should have returned two, something is wrong with the way you entered the information. Be careful not to blindly accept what the computer

tells you. Always question the output from the machine to make sure that it makes sense.

Before continuing your study of polynomial and rational functions, become familiar with the computer algebra system, or CAS, that you will be using throughout the rest of this course.

Task 3-3: Becoming Familiar with Your CAS

Work through the course handout for your CAS *interactively*. Don't just read about what your CAS can do. Try it! Experiment. Take notes. Try to understand why the syntax is the way it is, rather than simply memorizing the format for the various commands. You will probably make lots of errors at first. Most people do. Be patient. You'll get the hang of it. Promise.

In particular, review how to do the following types of tasks.

1. Symbolic calculations: Expand an expression. Factor. Simplify.
2. Numeric calculations: Evaluate an expression. Find its zeros. Express the result using a designated number of decimal places.
3. Function definitions: Use the built-in functions. Define your own functions. Represent a function using an expression, a table or a set of ordered pairs, and a piecewise-defined function.
4. Graphical calculations: Plot one or more functions on the same pair of axes. "Zoom in" or "zoom out" on a specified region. Determine how your CAS handles points of discontinuity: jumps, holes, and vertical asymptotes. Plot a table or a set of ordered pairs using a scatter plot.

Task 3-4: Using Your CAS to Examine the Behavior of Polynomial and Rational Functions

1. Use your CAS to inspect the following polynomial expressions. In each case:
   - **i.** Define the function.
   - **ii.** Find the x-intercepts (if any) where the graph crosses the horizontal axis. Enter the results in your Activity Guide.
   - **iii.** Find the y-intercept where the graph crosses the vertical axis. Enter the result in your Activity Guide.
   - **iv.** Plot the function. Be sure and include any "interesting" points—that is, intercepts or turning points—in your graph.
   - **v.** Print a copy of the graph and place it in your Activity Guide.

On your printout:

i. Label the intervals where the function is increasing and decreasing.

ii. Label the intervals where the function is concave up and down.

iii. Label the local maxima and minima.

a. $f(x) = -6x^2 - x + 15$

x-intercepts:

y-intercept:

Graph of function. Label the intervals where the function is increasing/decreasing and concave up/down. Label the local maxima/minima.

b. $g(x) = 90x^3 - 249x^2 + 168x$

x-intercepts:

y-intercept:

Graph of function. Label the intervals where the function is increasing/decreasing and concave up/down. Label the local maxima/minima.

c. $h(x) = x^4 - 8x^3 + 22x^2 - 24x + 8$

x-intercepts:

y-intercept:

Graph of function. Label the intervals where the function is increasing/decreasing and concave up/down. Label the local maxima/minima.

d. $r(x) = 2x^4 - 5x^3 + 10x - 12$

x-intercepts:

y-intercept:

Graph of function. Label the intervals where the function is increasing/decreasing and concave up/down. Label the local maxima/minima.

2. Use your CAS to inspect the following rational expressions. In each case:

i. Define the function.

ii. Find the factors of the numerator and the denominator. Do not simplify. Enter the results in your Activity Guide.

iii. Using the information from part ii, find the x-coordinates of any holes. Enter the results in your Activity Guide.

iv. Using the information from part ii, find any vertical asymptotes. Enter the results in your Activity Guide.

v. Using the information from part ii, find any x-intercepts. Enter the results in your Activity Guide.

vi. Find the y-intercept, if it exists. Enter the result in your Activity Guide.

vii. Plot the function. Be sure and include any "interesting" points—that is, intercepts, turning points, and asymptotes—in your graph. If it is not clear what is happening near an x-intercept or a turning point, zoom in and examine the graph more closely.

viii. Determine the domain of the function. Enter the information in your Activity Guide.

ix. Print a copy of the graph and place it in your Activity Guide.

On your printout:

i. Label the intervals where the function is increasing and decreasing.

ii. Label the intervals where the function is concave up and down.

iii. Label any local maxima and minima.

iv. Label any holes.

v. Label any vertical asymptotes.

a. $m(x) = \dfrac{2x^2 - 3x}{x^3 - 9x}$

Function factored:

x-coordinate of holes (if any):

Vertical asymptotes (if any):

x-intercepts (if any):

y-intercept (if exists):

Domain:

Graph of function. Label the intervals where the function is increasing/decreasing and concave up/down. Label the local maxima/minima, holes, and vertical asymptotes.

b. $t(x) = \dfrac{x^2 - 1}{x + 2}$

Function factored:

x-coordinate of holes (if any):

Vertical asymptotes (if any):

x-intercepts (if any):

y-intercept (if exists):

Domain:

Graph of function. Label the intervals where the function is increasing/decreasing and concave up/down. Label the local maxima/minima, holes, and vertical asymptotes.

3

c. $n(x) = \dfrac{x^2 - 2x - 8}{x^2 - 2x}$

Function factored:

x-coordinate of holes (if any):

Vertical asymptotes (if any):

x-intercepts (if any):

y-intercept (if exists):

Domain:

Graph of function. Label the intervals where the function is increasing/decreasing and concave up/down. Label the local maxima/minima, holes, and vertical asymptotes.

Unit 3 Homework After Section 1

- Complete the tasks in Section One in the Activity Guide. Be prepared to discuss them in class.
- Use the rules given below to do the skill exercise HW3.1 Try to identify any areas where you need extra help.

Rules for simplifying exponents.

$a^0 = 1$ $\quad (a^n)^m = a^{nm}$ $\quad (ab)^n = a^n b^n$

$a^{-n} = \dfrac{1}{a^n}$ $\quad a^n a^m = a^{n+m}$ $\quad \left(\dfrac{a}{b}\right)^n = \dfrac{a^n}{b^n}$

$a^{1/n} = \sqrt[n]{a}$ $\quad a^{m/n} = \sqrt[n]{a^m}$

Formula for finding roots of a quadratic equation:

If $ax^2 + bx + c = 0$, where $a \neq 0$, then

$$x = \frac{-b \pm \sqrt{b^2 - 4ac}}{2a}$$

HW3.1 Do some skill problems.

1. Evaluate the following arithmetic expressions.

a. $(4)^{-1/2}$

b. $\dfrac{(3)^{-1} + (3)^2}{2}$

c. $\dfrac{(8)^{5/3}}{(16)^{3/4}}$

f. $\dfrac{(-1)^3 + (3)^0}{-4}$

d. $(2^2)^3$

g. $\left(\dfrac{1}{9}\right)^{-2} + \left(\dfrac{1}{9}\right)^2$

e. $(6)^0$

h. $(8)^{-1/3}(16^{1/4} + 2^0)$

2. Simplify the following algebraic expressions.

a. $(x^3)^2$

e. $x^2 x^0$

b. $\dfrac{x^9}{x^7}$

f. $\sqrt{4x^0}$

c. $\dfrac{x^5 - x^3}{x^2}$

g. $\dfrac{x^3}{x^8} + \dfrac{2x^{16}}{x^{21}}$

d. $x^4 x^5$

h. $\dfrac{\sqrt{x}}{x^3} + \dfrac{x}{\sqrt{x^7}} + 3x$

3. Evaluate the following functions at the designated value.

a. Let $f(x) = x^{-1/3}x^2$. Evaluate f at $x = 8$.

b. Let $f(x) = 3\left(\dfrac{8 - x^2}{4}\right)^2$. Evaluate f at $x = 4$.

c. Let $f(x) = (x + 6)^2 x^{-4}$. Evaluate f at $x = 3$.

3

d. Let $f(x) = \dfrac{x^3}{6 - x^2}$. Find $f(2)$.

e. Let $f(x) = (2x)^{5/4}(x - 6)^{-6}$. Find $f(8)$.

f. Let $f(x) = (x + 8)^{-1/2}(x + 3)^2$. Find $f(1)$.

4. Consider $f(x) = 3x^2 - 2x + 9$. Suppose a is a real number in the domain of f. Suppose h is a very small real number.

a. Find an expression for

(1) $f(a)$ **(3)** $\dfrac{f(a + h) - f(a)}{h}$

(2) $f(a + h)$

b. Suppose $a = 2$ and $h = 0.1$. Find the value of

(1) $f(a + h)$ **(2)** $\dfrac{f(a + h) - f(a)}{h}$

c. Explain why $f(a + h) \neq f(a) + f(h)$. Support your explanation with an example.

5. Solve the following equations for x.

a. $5x^2 = 125$ **c.** $x^{-1/3}x = 4$

b. $(x - 1)^{-1/2} = \frac{1}{4}$

6. Find the roots of the following polynomials.

a. $x^3 - 4x$ **h.** $x^4 - 6x^3 + 9x^2$

b. $x^2 + 3x - 4$ **i.** $x^2 - 9$

c. $x^2 + 4x + 2$ **j.** $x^2 + 3x + 4$

d. $-x^2 + 3x + 6$ **k.** $x^2 + 3x - 4$

e. $-x^2 - 6x - 9$ **l.** $x^2 - x - 4$

f. $x^2 + 3x - 4$ **m.** $3x^2 - x + 2$

g. $2x^2 - 8x + 8$ **n.** $-2x^2 - 3x - 1$

7. Assume the following expressions represent rational functions.

i. Factor the numerator and denominator of the expression.

ii. Find the domain of the function.

iii. For each zero in the denominator of the expression, determine if the function has a hole or a vertical asymptote at the zero.

iv. Represent the rational function by a simplified expression and state its domain.

a. $\dfrac{x^2 + x - 2}{x - 1}$

b. $\dfrac{3x^3 + 6x^2 - 45x}{x - 3}$

c. $\dfrac{x^2 + 7x - 18}{x + 9}$

d. $\dfrac{x^2 + x - 2}{x - 1}$

e. $\dfrac{x^2 - 1}{x^2 + 5x + 4}$

f. $\dfrac{x^2 + 8x + 12}{x^2 - 4}$

8. Explain why $(a + b)^n \neq a^n + b^n$, for $n \neq 1$. Support your explanation with an example.

- Sketch a polynomial or a rational function satisfying the specified conditions in HW3.2.

HW3.2 For each of the following sets of conditions:

i. Sketch the graph of a function that satisfies the given conditions.

ii. Represent a function satisfying the given conditions by an expression.

1. A polynomial of degree 1 with slope 2 and y-intercept -2.5.

2. A polynomial of degree 2 with a turning point at $P(6.5,2.25)$, which is decreasing on the open interval $(-\infty,6.5)$, and increasing on the open interval $(6.5,\infty)$.

3. A quadratic function f, where $f(x)$ is negative for all values of x.

4. A quadratic function with no real roots that is concave up.

5. A quadratic function f with exactly one root at $x = 1.75$ such that $f(x) \geq 0$ for all values of x.

6. A linear function whose range consists of a single value, 4.

7. A one-to-one, cubic function whose domain and range is the set of all real numbers.

8. A linear function whose output for a given input is equal to three times the value of the particular input minus 12.5.

9. A rational function with holes at $x = -2$ and $x = 4$ and no vertical asymptotes.

10. A rational function with a vertical asymptote at $x = 0$.

11. A rational function that intersects the horizontal axis at $x = 2$ and $x = 4$, has a hole at $x = 1$, and a vertical asymptote at $x = 3$.

12. A quadratic function with roots 0.9 and -0.5.

13. Two distinct quadratic functions that have the same two real roots.

14. Two distinct linear functions that are both decreasing and have the same slope.

- Examine how absolute value functions and some other basic functions behave in HW3.3 and HW3.4.

HW3.3 Examine some absolute value functions.

1. Find:

a. $|23.6|$

b. $|-23.6|$

c. $|0|$

d. $|x - 1|$ when $x = -3$

2. Describe how to calculate the absolute value of a given number.

3. Sketch the graphs of the following functions. Use your CAS to check your sketches.

a. $y = |x|$

b. $y = |x + 1|$

c. $y = |2x - 0.5|$

4. Represent each of the three functions in part 3 by a piecewise-defined function.

5. Describe the shape of the graph of $y = |mx + b|$, where $m \neq 0$.

6. Give the coordinates of the turning point of $y = |mx + b|$, where $m \neq 0$.

HW3.4 Examine some basic functions. For each of the basic functions listed below:

i. Find the domain of the function.

ii. Plot the function, using your CAS. Print a hard copy of your graph. Label any vertical asymptotes, holes, and local maxima/minima. Label the intervals where the function is increasing/decreasing and concave up/down.

iii. Find the range of the function.

1. Basic cubic: $f(x) = x^3$

2. Reciprocal function: $f(x) = \dfrac{1}{x}$ or x^{-1}

3. Square root function: $f(x) = \sqrt{x}$ or $x^{1/2}$

4. Half-circle: $f(x) = \sqrt{1 - x^2}$

- Use your CAS to investigate some questions in HW3.5.

HW3.5 Use your CAS to investigate the answers to the following questions.

1. Graph the following functions on the same pair of axes: $p(x) = \frac{1}{2}x^2$, $q(x) = x^2$, and $r(x) = 2x^2$. Print a copy of the graph. Label the functions. In general, how does the value of a affect the shape of the graph of $f(x) = ax^2$?

2. Graph the following functions on the same pair of axes: $p(x) = x^2 - 2$, $q(x) = x^2$, and $r(x) = x^2 + 2$. Print a copy of the graph. Label the functions. In general, how does the value of c affect the location of the graph of $f(x) = x^2 + c$?

3. Examine the values of the *even power function* $e(x) = x^4$ as x approaches $-\infty$ and as x approaches $+\infty$. Similarly, examine the values of the *odd power function* $o(x) = x^3$ as x approaches $-\infty$ and as x approaches $+\infty$. Print copies of the two graphs. Consider the general situation, $f(x) = x^n$, where n is a positive integer. How does the fact that n is an even integer versus n being an odd integer affect the shape of the graph of the power function $f(x) = x^n$?

4. Compare the growth of $f(x) = 50x^2$ to the growth of $g(x) = x^3$. Which grows faster for large positive values of x? Compare the growth of $f(x) = 100x^2$ to the growth of $g(x) = 0.1x^3$. Which grows faster for large positive values of x? In general, compare the growth of $f(x) = ax^2$ to the growth of $g(x) = bx^3$, where a and b are positive real numbers. Which grows faster for large positive values of x (regardless of the size of a and b)?

5. Consider the rational function

$$s(x) = \frac{x^2 + 2x}{x(x^2 - 0.25)}$$

Find all the vertical asymptotes and holes in the graph of s. Investigate the behavior of the function s near each asymptote (hole) by evaluating the function at 10 values close to the asymptote (hole)—with 5 values to the right and 5 to the left. For each case, exhibit the results in a table and explain what they tell you about the behavior of the graph of the function near that point.

3

SECTION 2

Trigonometric Functions

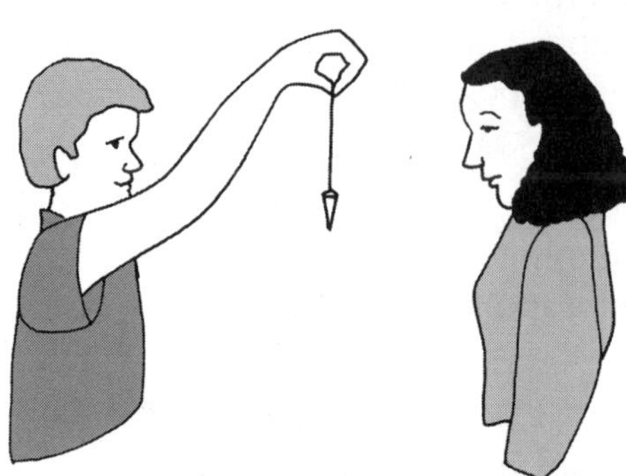

Functions can be used to model situations in the real world. Trigonometric functions are particularly useful in this regard, since they are periodic. For instance, trigonometric functions can be used to model the motion of a weight attached to a spring as it bobs up and down, the movement of sound waves and light waves, the oscillation of a pendulum, the length of the days and nights throughout the year, and a person's biorhythms.

The goal of this section is to analyze the behavior of the six fundamental trigonometric functions. What are the shapes of their graphs? What are their domains? Ranges? Where are they increasing? Decreasing? Concave up? Concave down? Where are their turning points? Zeros? Vertical asymptotes? How are they related to one another?

Before you explore the behavior of the trigonometric functions, you need to think about various ways of measuring the size of an angle. Just as there are several different ways to measure distance, for example using yards or meters, there are several different ways to measure the size of an angle. One way is to use *degree measure.* Another method is to use *radian measure.* In the next task, you will examine these two measures and practice converting between them.

First, a few comments about notation and the difference between radian and degree measure: Consider a unit circle—that is, a circle with radius 1 and center at the origin—and a vector of length 1, whose initial point is at the center of the circle and whose terminal point is on the perimeter of the circle. Assume the vector rotates around the circle, always starting on the positive x-axis. As the vector rotates, it forms an angle. The size of the angle is determined by how far the vector rotates with respect to the positive x-axis. The sign of the angle is determined by the direction the vector rotates; if the vector rotates in a counter-clockwise direction, the associated angle is positive, and if it rotates in a clockwise direction, the angle is negative. The question is: How can you measure the angle formed by the vector? One way is to use degrees. When the vector rotates all the way around the circle in the positive, or counterclockwise, direction, it forms a $360°$ angle. Consequently, if the vector rotates one-third of the way around the circle in the positive direction, it forms a $120°$ angle. If it rotates one-eighth of the way around in a clockwise direction, it forms a $-45°$ angle, and so on. (See Figure 1.)

The *radian* measure of an angle, on the other hand, is the length of the arc that the vector cuts from the unit circle with the appropriate sign attached—positive for counterclockwise and negative for clockwise. Since the circumference of a circle equals 2π times the radius of the circle, the circumference of the unit circle is $2\pi \cdot 1$ or 2π. Therefore, when the vector rotates all the way around the circle in a positive direction, it cuts an arc of length 2π, and the size of the resulting 360° angle is 2π radians; that is,

$$360° = 2\pi \text{ radians}$$

This relationship provides a handy way to convert back and forth between degree and radian measure. For example,

$$120° = \tfrac{1}{3}(2\pi) \text{ radians or } \tfrac{2}{3}\pi \text{ radians}$$

and

$$-45° = -\tfrac{1}{8}(2\pi) \text{ radians or } -\tfrac{\pi}{4} \text{ radians}$$

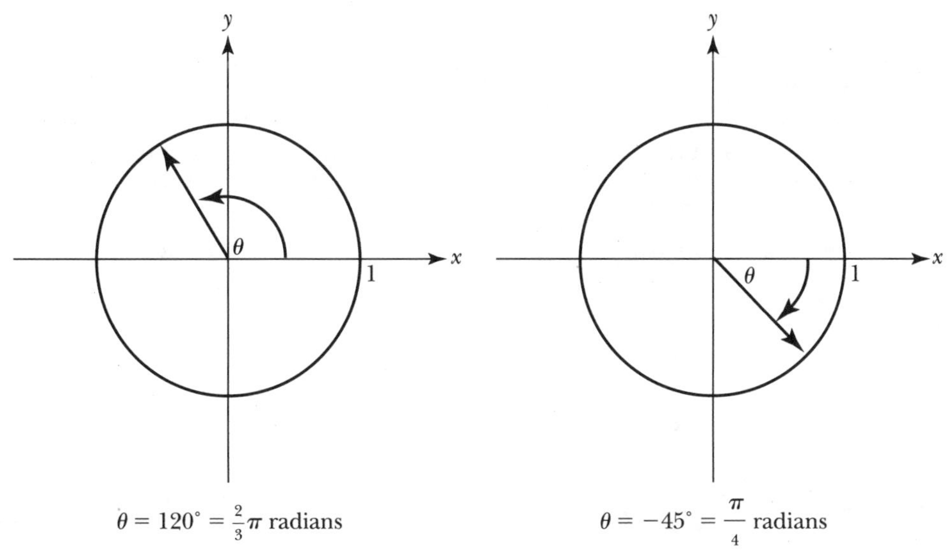

Figure 1.

Task 3-5: Measuring Angles

1. Convert between radian and degree measure.

a. Consider some positive angles. Using increments of 45°, label the circle below in degrees and radians, for $0° < \theta \leq 360°$. Indicate each angle with an arrow. (To help you get started, the first angle is already labeled.)

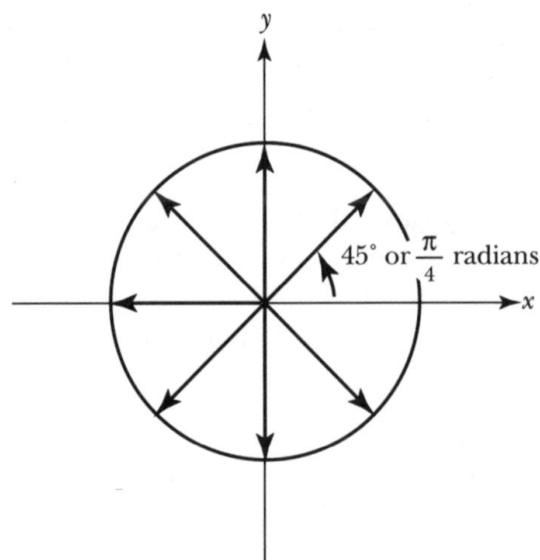

b. Consider some negative angles, by reversing direction and traveling clockwise around the unit circle. Using increments of $-30°$, label the circle below in degrees and in the equivalent value in radians in terms of π, for $-180° \leq \theta < 0°$. Indicate each angle with an arrow.

Note: The first angle is marked $-30°$ and with its equivalent value $-\frac{\pi}{6}$ radians.

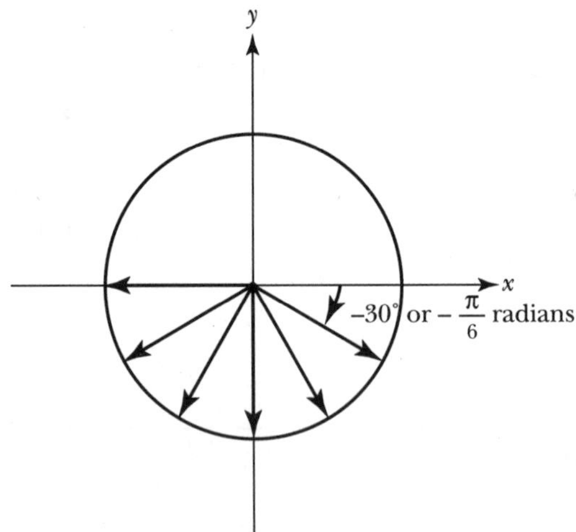

c. Approximate the decimal value of some radian measures.

When a CAS plots a trigonometric function, it usually labels the horizontal axis in terms of the decimal equivalent for radians using the fact that

$$180° = \pi \text{ radians} \approx 3.1416 \text{ radians}$$

To help you think about the axis being labeled in this manner, label all the $\frac{1}{4}$ values of π between -2π and 2π—that is, label the values of -2π, $-\frac{7}{4}\pi$, $-3/2\pi$, ..., 2π—on the number line given below.

Note: -2π*, which is approximately equal to* -6.28*, is already labeled.*

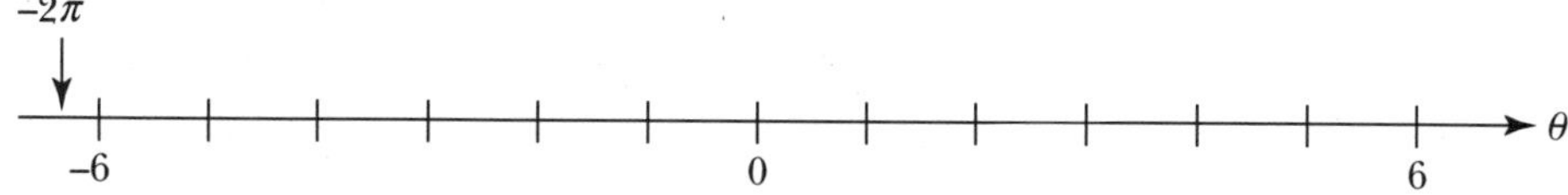

d. Practice converting back and forth.

Convert between degrees and radians (as a decimal value and in terms of π) as you fill in the missing values on the table given below.

Degrees	Radians (π)	≈Radians (#)
0		
	$\frac{\pi}{4}$	
		1.5708
135		
		3.1416
	$\frac{3\pi}{2}$	
360		
		9.4248
	4π	
−45		
	$-\frac{\pi}{2}$	
		−3.1416
	$-\frac{3\pi}{4}$	
−360		
		−12.5664

3

2. Examine the relationship between an angle and the terminal point of its associated vector.

As usual, suppose θ is an angle created by rotating a vector of length 1 around the unit circle. Let $P(x_0,y_0)$ be the terminal point of the vector.

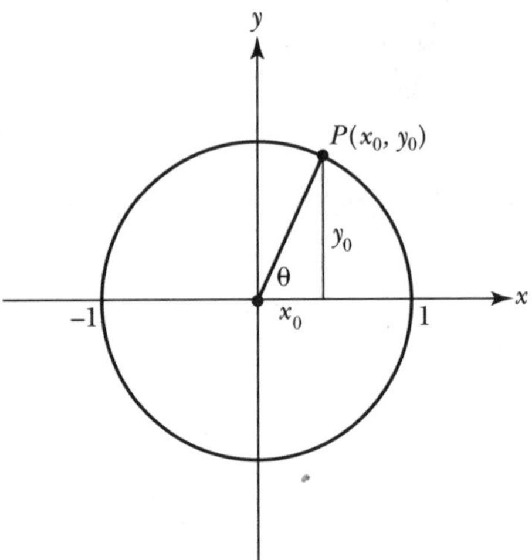

a. Find the coordinates of the terminal point $P(x,y)$ of the vector for given values of θ by filling in the table below.

θ	-2π	$-\frac{3\pi}{2}$	$-\pi$	$-\frac{\pi}{2}$	0	$\frac{\pi}{2}$	π	$\frac{3\pi}{2}$	2π
x									
y									

b. Find the values of the coordinates of the terminal point of the vector when $\theta = \frac{\pi}{4}$ radians.

(1) The Pythagorean Theorem gives a relationship between the lengths of the sides of a right triangle and its hypotenuse. State this relationship for the triangle given below.

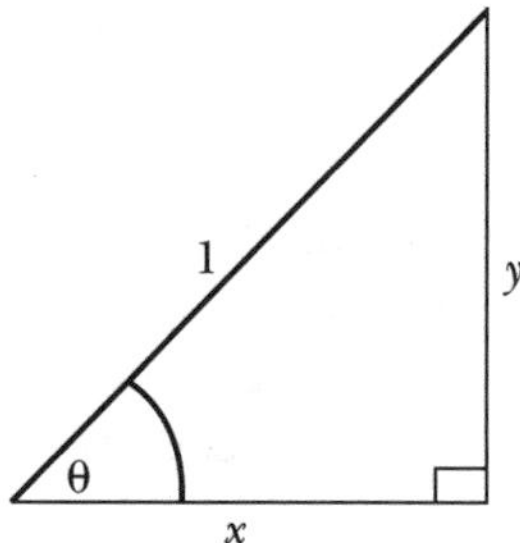

(2) If $\theta = \frac{\pi}{4}$ radians, the lengths of the two sides of the right triangle given above are equal. Find the length of each side. (Express the values in terms of a radical, not a decimal.)

c. The unit vector can rotate any number of times around the circle. The size of the angle reflects the number of rotations. For example, if the vector rotates around twice in a counterclockwise direction it forms an angle of 720° or 4π radians. If it rotates one and one-half times around in a clockwise direction, it creates an angle of size $-540°$ or -3π radians. In general, the size of the angle gets larger and larger as the vector rotates around the circle in the counterclockwise direction, and it gets smaller and smaller (more and more negative) as the vector rotates in the clockwise direction. As the vector repeatedly rotates around the circle (in either direction), the terminal point of the vector, $P(x,y)$, passes through the same points over and over again. As a result, many different angles—both positive and negative—are associated with the same values of x and y.

(1) Describe the set of all positive and negative values of θ where the x-coordinate of the terminal point of the associated vector is 0—that is, where $x = 0$.

(2) Describe the set of all positive and negative values of θ where the y-coordinate of the terminal point of the associated vector is 0—that is, where $y = 0$.

3

(3) Describe the set of all positive and negative values of θ where the x- and y-coordinates of the terminal point of the associated vector are equal—that is, where $x = y$.

So, you have a unit vector rotating around a unit circle. The vector forms an angle with the positive x-axis. When the vector rotates in a counterclockwise direction, the resulting angle is positive; otherwise it is negative. The angles can be measured in degrees or radians. Moreover, as the vector goes around and around the circle, its terminal point repeatedly passes through the same values, and the associated angle gets more and more positive or more and more negative.

The domains of the various trigonometric functions are collections of angles. The value of a particular trigonometric function at a given angle is determined by the coordinates of the terminal point of the associated vector. In other words, a trig function accepts an angle as input and then returns a number that is derived from the coordinates of the terminal point of the associated vector. In particular, if θ is the angle formed by rotating a vector around the unit circle and $P(x,y)$ is the point where the vector terminates on the circle, the *cosine* function returns the first coordinate of P, the *sine* function returns the second coordinate, the *tangent* function returns the quotient of the coordinates, and the *secant, cosecant,* and *cotangent* functions are reciprocals of these three basic functions; that is, the six fundamental trigonometric functions are defined as follows:

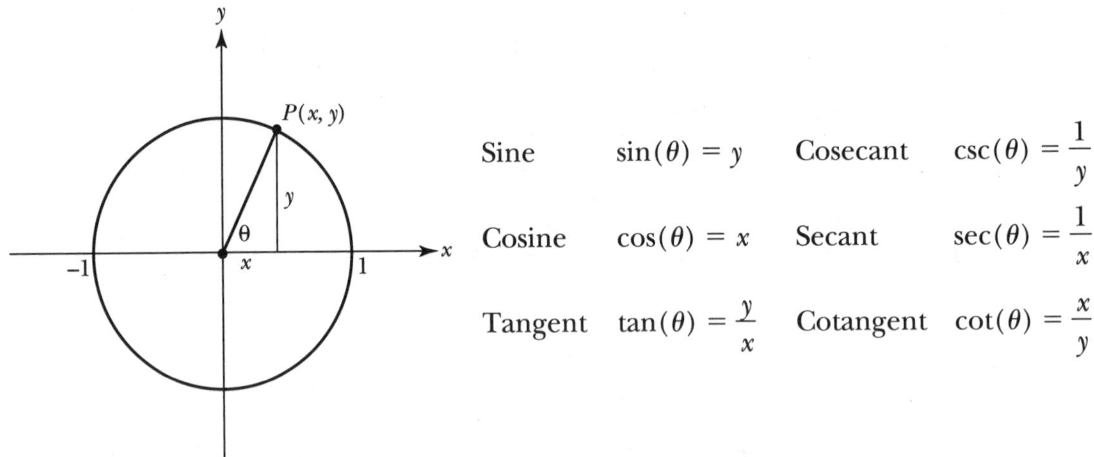

Note: If you are accustomed to defining the trigonometric functions using a right triangle, observe that the length of the hypotenuse of the triangle is always equal to 1 with the unit circle approach. Consequently, $\sin(\theta) = y$*, or*

$\sin(\theta) = y/1$, is equivalent to $\sin(\theta) = y/h$, where h is the length of the hypotenuse of the corresponding triangle.

Since the terminal point of the vector repeatedly passes through the same points on the unit circle, and since the values of the trigonometric functions depend on the coordinates of these points, the values of each function are also repetitive. As a result, the trigonometric functions are said to be *periodic* functions, where the period of the function is the smallest interval containing one complete cycle through the sequence of values.

In the next task, you will investigate the trigonometric functions. You will evaluate them at various angles, find their domains, and analyze their graphs.

Task 3-6: Graphing Trigonometric Functions

1. Describe the domains of the trigonometric functions. For each function, describe the values of θ where the function is *not* defined.

Note: Some trig functions are defined for any value of θ. Others have restrictions on their domains, since you cannot divide by 0. Use the definitions of the trigonometric functions and the results from the end of Task 3.5 to determine which functions have restrictions on their domains and the values where these functions are not defined.

a. $\sin(\theta)$

b. $\cos(\theta)$

c. $\tan(\theta)$

d. $\csc(\theta)$

e. $\sec(\theta)$

f. $\cot(\theta)$

2. Find the places where the graphs of the trigonometric functions intersect the horizontal axis; that is, for each of the trigonometric functions, describe the set of all values of θ where the function is equal to zero.

Note: Use the definitions of the trigonometric functions, the results from the end of Task 3-5, and the fact that the quotient of two numbers equals 0 if and only if the numerator of the quotient equals 0.

a. $\sin(\theta)$

b. $\cos(\theta)$

3

c. $\tan(\theta)$

d. $\csc(\theta)$

e. $\sec(\theta)$

f. $\cot(\theta)$

3. Sine, cosine and tangent are considered to be the basic trigonometric functions since the other three functions can be expressed in terms of them.

a. Express $\csc(\theta)$ in terms of $\sin(\theta)$.

$\csc(\theta) =$

b. Express $\sec(\theta)$ in terms of $\cos(\theta)$.

$\sec(\theta) =$

c. Express $\cot(\theta)$ in terms of $\tan(\theta)$.

$\cot(\theta) =$

d. Actually you only need the sine and cosine functions, since the tangent function, and hence the cotangent function, can be expressed in terms of these functions.

(1) Express $\tan(\theta)$ in terms of $\sin(\theta)$ and $\cos(\theta)$.

$\tan(\theta) =$

(2) Express $\cot(\theta)$ in terms of $\sin(\theta)$ and $\cos(\theta)$.

$\cot(\theta) =$

4. Evaluate the trigonometric functions at various angles.

Fill in the table given below using the definitions (which are re-stated for you). Determine the entries in an efficient manner. First find all the values in the x and y columns. Use this information to quickly fill in the sine column. Use the entries in the sine column, and the fact that

$$\csc(\theta) = \frac{1}{\sin(\theta)}$$

to quickly fill in the entries in the cosecant column, and so on.

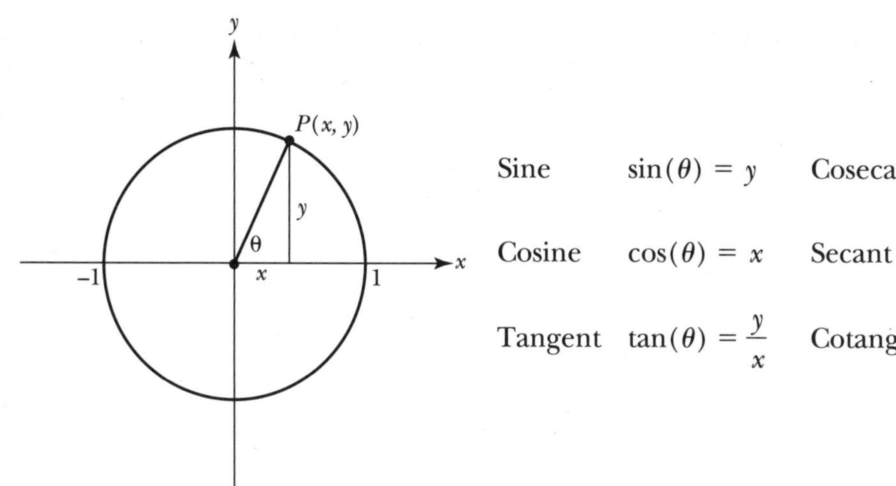

Sine	$\sin(\theta) = y$	Cosecant	$\csc(\theta) = \dfrac{1}{y}$
Cosine	$\cos(\theta) = x$	Secant	$\sec(\theta) = \dfrac{1}{x}$
Tangent	$\tan(\theta) = \dfrac{y}{x}$	Cotangent	$\cot(\theta) = \dfrac{x}{y}$

θ	x	y	$\sin(\theta)$	$\csc(\theta)$	$\cos(\theta)$	$\sec(\theta)$	$\tan(\theta)$	$\cot(\theta)$
0								
$\frac{\pi}{4}$								
$\frac{\pi}{2}$								
$\frac{3\pi}{4}$								
π								
$\frac{3\pi}{2}$								
2π								
$-\frac{\pi}{4}$								
$-\frac{\pi}{2}$								
$-\frac{3\pi}{4}$								
$-\pi$								

5. Graph the trigonometric functions.

Note: Since your CAS may not understand the symbol θ, we have replaced the variable name θ with t.

a. Use your CAS to graph the six trigonometric functions, for $-2\pi \leq t \leq 2\pi$. Print a copy of the graph. Label the graphs $\sin(t)$, $\cos(t)$, and so on, and place them in below.

b. Relabel the horizontal axis of each graph in terms of π.

6. Analyze the graphs of the three basic trigonometric functions.

Use your graphs from part 5 to fill in the chart given below, for $-2\pi \leq t \leq 2\pi$. Express your responses in terms of π. For example, to enter the zeros for $\sin(t)$, write -2π, $-\pi$, 0, π, 2π. To indicate the intervals where $\sin(t)$ is increasing, use interval notation and write $(-2\pi, -\frac{3\pi}{2})$, $(-\frac{\pi}{2}, \frac{\pi}{2})$, $(\frac{3\pi}{2}, 2\pi)$.

Function	sin(*t*)	cos(*t*)	tan(*t*)
Range			
Zeros			
Location of vertical asymptotes			
Intervals where increasing			
Intervals where decreasing			
Intervals where concave up			
Intervals where concave down			

Most real-life situations, such as the length of the days and nights throughout the year, cannot be modeled by the basic trigonometric functions. Usually the function has to be stretched or shrunk vertically or horizontally, and shifted left, right, up, or down to fit the data under consideration. In the case of the sine function, this can lead to messy looking functions such as

$$f(t) = 0.25 \sin(4t - 6) + 10.99$$

or

$$f(t) = 0.25 \sin(4(t - \tfrac{3}{2})) + 10.99$$

3

where the *general sine function* has the form

$$f(t) = a \sin(b(t - c)) + d$$

The values of a, b, c, and d are real numbers, which determine the shape of the function's graph. They indicate how the graph of the general sine function can be obtained by shrinking, stretching, and shifting the graph of $\sin(t)$.

In the next two tasks, you will use your CAS to determine how varying the values of a, b, c and d affects the shape and location of the basic sine function. To begin with, you will change one coefficient at a time, and then in the homework you will examine the impact of changing all of them at once.

Task 3-7: Stretching and Shrinking the Sine Function

1. Examine the impact of the value of a by comparing the graph of $\sin(t)$ to the graph of $f(t) = a\sin(t)$, when $a > 0$.

a. Consider some specific functions. For each of the functions given below in (1)–(4):

i. Use your CAS to graph the given function and $\sin(t)$ on the same pair of axes for $-2\pi \leq t \leq 2\pi$. Visually compare the shapes of the two graphs.

ii. Print a copy of the graphs of the two functions and place it in your Activity Guide. Label the given function and $\sin(t)$.

iii. State the range of the given function.

(1) $f(t) = 4\sin(t)$

Range of $f(t) = a\sin(t)$, when $a = 4$:

(2) $f(t) = 2.5\sin(t)$

Range of $f(t) = a\sin(t)$, when $a = 2.5$:

(3) $f(t) = \frac{1}{2}\sin(t)$

Range of $f(t) = a\sin(t)$, when $a = \frac{1}{2}$:

(4) $f(t) = 0.25\sin(t)$

Range of $f(t) = a\sin(t)$, when $a = 0.25$:

b. Generalize your observations.

(1) Describe the impact the value of a has on the graph of $f(t) = a\sin(t)$ by comparing the shape of the graph to the graph of $\sin(t)$.

(a) Describe the impact of a when $0 < a < 1$.

(b) Describe the impact of a when $a > 1$.

(2) Give the range of $f(t) = a\sin(t)$ in terms of a.

2. Analyze the impact of the value of b by comparing the graph of $\sin(t)$ to the graph of $f(t) = \sin(bt)$, where $b > 0$.

a. For each of the functions given below in (1)–(4):

i. Use your CAS to graph the given function and $\sin(t)$ on the same pair of axes over the specified interval. Visually compare the shapes of the two graphs.

ii. Print a copy of the graphs of the two functions and place it in your Activity Guide. Label the given function and $\sin(t)$. Relabel the horizontal axis in terms of π.

iii. Determine the period of the given function in terms of π.

Note: In the case of $f(t) = \sin(t)$, $b = 1$ since $\sin(t) = \sin(1 \cdot t)$. Moreover, the period of $\sin(t)$ is $\frac{2\pi}{1}$ or 2π, since the graph of $\sin(t)$ repeats itself after one rotation (2π radians) around the circle.

(1) $f(t) = \sin(2t)$, where $-2\pi \le t \le 2\pi$

Period of $f(t) = \sin(bt)$, when $b = 2$:

(2) $f(t) = \sin(4t)$, where $-2\pi \leq t \leq 2\pi$

Period of $f(t) = \sin(bt)$, when $b = 4$:

(3) $f(t) = \sin(t/2)$, where $-4\pi \leq t \leq 4\pi$

3

Period of $f(t) = \sin(bt)$, when $b = \frac{1}{2}$:

(4) $f(t) = \sin(0.25t)$, where $-8\pi \leq t \leq 8\pi$

Period of $f(t) = \sin(bt)$, when $b = 0.25$:

b. Generalize your observations.

(1) Describe the impact of the value of b on the graph $f(t) = \sin(bt)$ by comparing the shape of the graph to the graph of $\sin(t)$.

(a) Describe what happens when $0 < b \leq 1$.

(b) Describe what happens when $b > 1$.

(2) Give the period of $f(t) = \sin(bt)$ in terms of b.

In the last task, you used your CAS to examine how the values of the coefficients a and b affect the shape of the graph of the general sine function $f(t) = a\sin(b(t - c)) + d$. In the next task, you will use a similar approach to investigate how the values of c and d affect the location of the graph.

Task 3-8: Shifting the Sine Function

1. Investigate the impact of the value of c by comparing the graph of $\sin(t)$ to the graph of $f(t) = \sin(t - c)$.

   a. Use your CAS to explore how the value of c affects the location of the graph of $f(t) = \sin(t - c)$. Examine functions where c is positive and where c is negative by graphing functions such as

$$f(t) = \sin(t - \pi/2)$$

$$f(t) = \sin(t + 1)$$

$$f(t) = \sin(t + \pi)$$

for $-2\pi \leq t \leq 2\pi$, on the same pair of axes with $\sin(t)$. Make copies of your graphs and place them in the space below. Label the functions. Relabel the horizontal axis in terms of π.

b. Generalize your observations.

Describe the impact of the value of c on the graph of $f(t) = \sin(t - c)$ by comparing the location of the graph to the graph of $\sin(t)$. Describe what happens when $c < 0$ and when $c > 0$.

2. Analyze the impact of the value of d by comparing the graph of $\sin(t)$ to graph of $f(t) = \sin(t) + d$, where $d \neq 0$.

   a. Use your CAS to discover how the value of d affects the graph of $f(t) = \sin(t) + d$. Examine functions where d is positive and where d is negative by graphing functions such as

$$f(t) = \sin(t) + 4$$

$$f(t) = \sin(t) - 2$$

$$f(t) = \sin(t) + 0.5$$

   for $-2\pi \leq t \leq 2\pi$, on the same pair of axes with $\sin(t)$. Make copies of your graphs and place them in the space below. Label the functions. Give the range of each function.

b. Generalize your observations.

(1) Describe the impact the value of d has on the graph of $f(t) = \sin(t) + d$ by comparing the location of the graph to the graph of $\sin(t)$. State what happens when $d < 0$ and when $d > 0$.

(2) Give the range of $f(t) = \sin(t) + d$ in terms of d.

Unit 3 Homework After Section 2

- Complete the tasks in Section Two in the Activity Guide. Be prepared to discuss them in class.
- Analyze the shape of some general sine functions in HW3.6.

HW3.6 Consider the general sine function $f(t) = a\sin(b(t - c)) + d$.

1. Summarize the results of Tasks 3.7 and 3.8.

Value	Impact on the Graph of sin(t)
$0 < a < 1$	
$a > 1$	
$0 < b < 1$	
$b > 1$	
$c < 0$	
$c > 0$	
$d < 0$	
$d > 0$	

3

2. For each of the following sinusoidal functions:
   i. Compare the shape of the graph to the graph of $\sin(t)$. Analyze:
   - horizontal stretch or shrink
   - vertical stretch or shrink
   - shift up or down
   - shift left or right

   ii. Sketch the graph by hand. Carefully label the axes on your graph.
   iii. Use your CAS to check that your sketch is correct.

   a. $f(t) = \sin(2t) + 6$
   b. $f(t) = 2\sin(t - \pi)$
   c. $f(t) = \sin(\frac{\pi}{5}(t - 5)) - 3$

3. Represent each of the following graphs by an expression of the form $a\sin(b(t - c)) + d$.
   i. Find the size of the shift up or down. Use this information to determine the value d.
   ii. Find the range of the function. Use this information to determine the value a.
   iii. Find the period of the function. Use this information to determine the value b.
   iv. Find the size of the shift left or right. Use this information to determine the value c.

a.

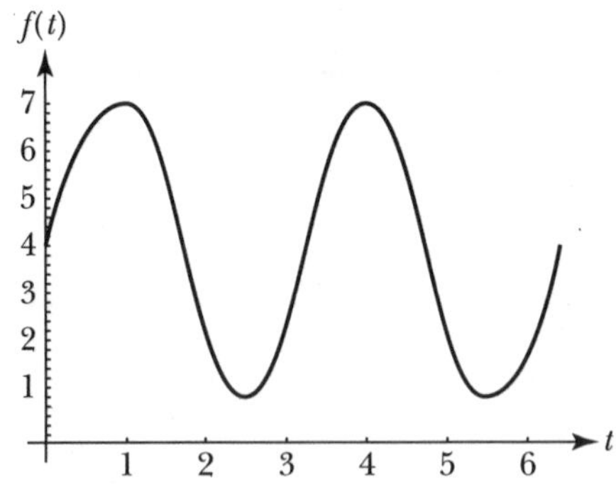

b.

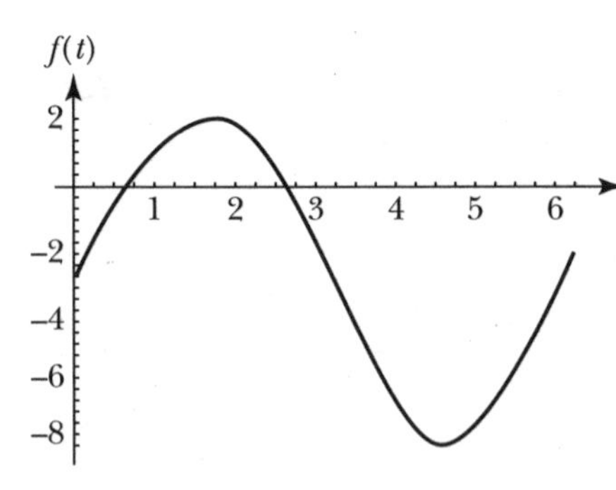

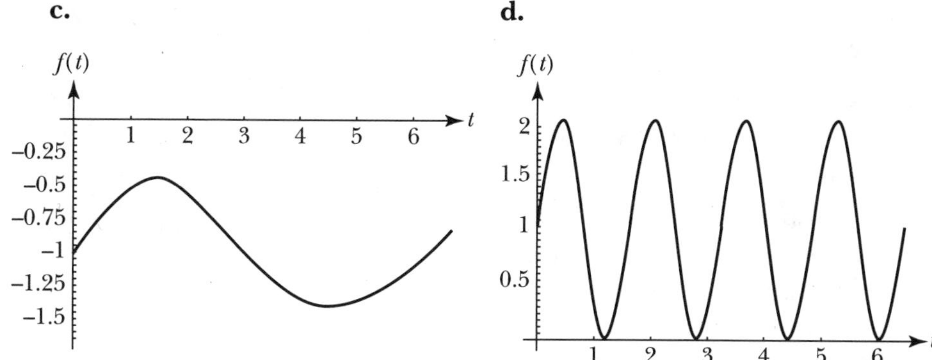

- Express the lengths of days and nights as general sine functions in HW3.7.

HW3.7 The graph below displays the day's length throughout 1992 for latitude 41°N. Note that an equinox—a day when the hours of daylight and darkness are nearly equal—occurs each spring and fall when the sun shines directly at the equator. The summer solstice—the longest day of the year—occurs when the earth tilts the northern hemisphere closest to the sun, and the winter solstice—the longest night of the year—occurs when the southern hemisphere is closest to the sun.

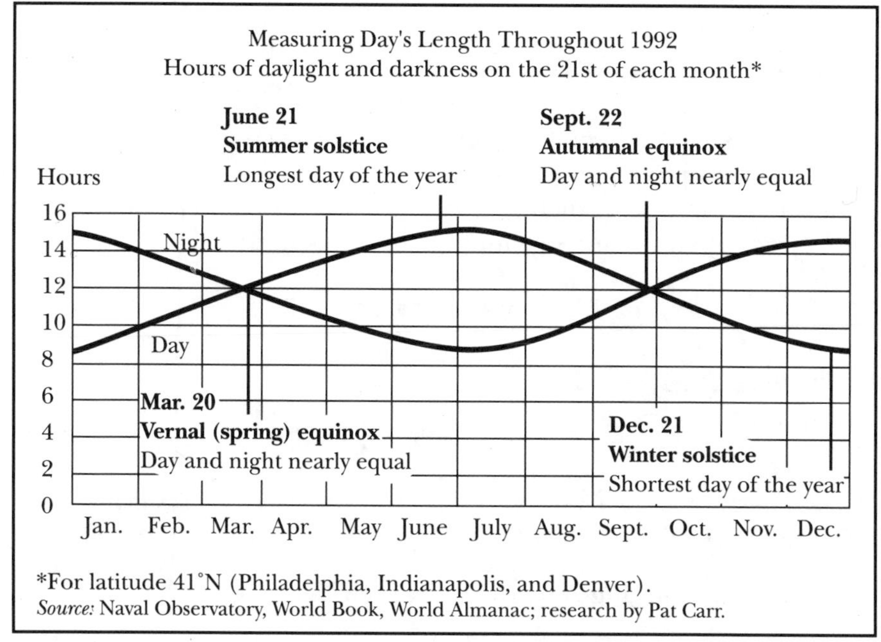

*For latitude 41°N (Philadelphia, Indianapolis, and Denver).
Source: Naval Observatory, World Book, World Almanac; research by Pat Carr.

1. Write a paragraph detailing what the graph tells you about the relationship between the length of the days and nights over the course of a year.

2. Represent the length of the days throughout the year by a general sine function called *day*.

   a. Assume $day(t) = a\sin(b(t - c)) + d$. Suppose the vernal equinox corresponds to the origin in the basic sine function. Find the values of d and a.

   b. Denote the first day of the successive months by 0 to 11—that is, denote January 1 by 0, February 1 by 1, and so on. How would you denote the end of December? What is the period of function *day*? Use this information to find the value of b.

   c. Assume each month has 30 days. Given that March 1 is denoted by 2, how would you denote the vernal equinox, which occurs on March 20? Determine the value of c.

   d. Use your CAS to check that your function is correct. Set the range of the graph to be 0 to 16 so your graph looks similar to the one given above.

3. Represent the length of the nights throughout the year by a general sine function called *night* by modifying the general sine function for *day*.

4. Plot *night* and *day* on the same pair of axes using your CAS. Print a copy of the graphs.

- Examine a concise way of expressing multiples of π in HW3.8.

HW3.8 The six fundamental trigonometric functions are periodic, since their values keep repeating themselves. For example, $\sin(t)$ is 0 when t is *any* integer multiple of π, such as -10π, 0π, 8π, 23π, and so on. A concise way to represent all the integer multiples of π is to write:

$$n\pi, \text{ where } n = 0, \pm 1, \pm 2, \pm 3, \ldots$$

which stands for

$$\ldots, -3\pi, -2\pi, -\pi, 0, \pi, 2\pi, 3\pi, \ldots$$

Note: The sequence of three dots on either end of the list implies that the list continues on indefinitely using the same pattern.

1. Consider the following sequences. In each case:

   i. Give a verbal description of all the multiples of π represented by the notation.

   ii. Give the sequence for which the notation stands.

   a. $2n\pi$, where $n = 0, \pm 1, \pm 2, \pm 3, \ldots$

   b. $(2n + 1)\pi$, where $n = 0, \pm 1, \pm 2, \pm 3, \ldots$

c. $\frac{n\pi}{2}$, where $n = 0, \pm 1, \pm 2, \pm 3, ...$

d. $\frac{(2n+1)\pi}{2}$, where $n = 0, \pm 1, \pm 2, \pm 3, ...$

2. Express each of the following sequences as a multiple of π.

a. $..., -8\pi, -4\pi, 0, 4\pi, 8\pi, ...$

b. $...-\pi, -\frac{3\pi}{4}, -\frac{\pi}{2}, -\frac{\pi}{4}, 0, \frac{\pi}{4}, \frac{\pi}{2}, \frac{3\pi}{4}, \pi, ...$

3. Express the answer to each of the following as a sequence or as a multiple of π. Support your answer with a sketch.

a. Find all the values of θ such that $\cos(\theta) = -1$.

b. Find all the values of θ such that $\sin(\theta) = -1$.

c. Find all the values of θ such that $\sin\theta = \cos\theta$.

d. Find all the values of θ such that $\sin\theta + \cos\theta = 0$.

- Consider some fundamental trigonometric identities in HW3.9.

HW3.9 Show that the following identities hold.

1. Use the Pythagorean Theorem and the definitions of sine and cosine to show that

$$\sin^2 \theta + \cos^2 \theta = 1$$

Note: $\sin^2 \theta = (\sin \theta)^2$, $\cos^2 \theta = (\cos \theta)^2$, *and so on.*

2. Use the Pythagorean Theorem and the definitions of tangent and secant to show that

$$\tan^2 \theta + 1 = \sec^2 \theta$$

3. Explain why $\sin(\theta + 2\pi) = \sin(\theta)$, for every value of θ.

4. Explain why $\cos(\theta + 2\pi) = \cos\theta$, for every value of θ.

5. Explain why $\sin(\theta + \pi) = -\sin(\theta)$, for every value of θ.

6. Explain why $\cos(\theta + \pi) = -\cos(\theta)$, for every value of θ.

- Graph some interesting combinations of functions in HW3.10.

HW3.10 For each of the following combination of functions:

i. Try to predict the shape of the graph of the combination.

ii. Check your prediction using your CAS.

iii. Print a copy of the graph.

1. $f(t) = -\sin(t)$, where $-4\pi \leq t \leq 4\pi$.

2. $f(t) = \sin(t) + t$, where $-4\pi \leq t \leq 4\pi$.

3. $f(t) = \sin(t) - t$, where $-4\pi \leq t \leq 4\pi$.

4. $f(t) = t\sin(t)$, where $-4\pi \leq t \leq 4\pi$.

5. $f(t) = t^2 \sin(t)$, where $-4\pi \leq t \leq 4\pi$.

6. $f(t) = (\sin(t))/t$, where $-4\pi \leq t \leq 4\pi$ but $t \neq 0$. What type of discontinuity does f appear to have at $t = 0$?

SECTION 3

Exponential and Logarithmic Functions

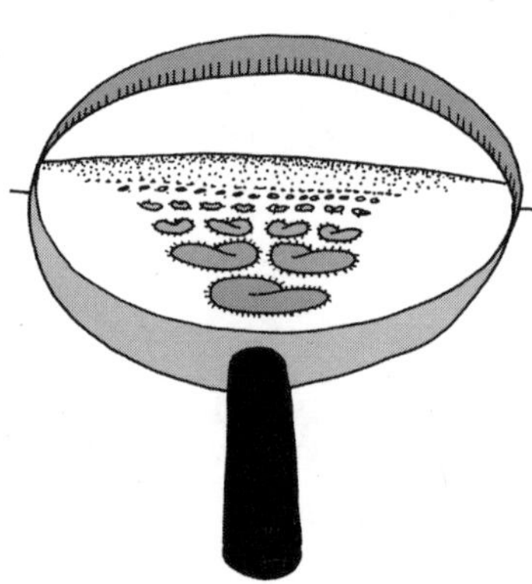

Suppose your best friend is a biologist who studies the growth rate of paramecia. Currently, she is examining a variety where every hour the size of the total population doubles as each paramecium becomes two paramecia. In this case, the population is said to be "growing exponentially," and with this rapid rate of growth, it will not take long for her to have a lot of paramecia. In the next task you will use an exponential function to model your friend's growth situation and investigate the impact on the model of changing the initial condition. You will also model the decay situation that results when her paramecia suddenly begin to die.

Task 3-9: Modeling Situations Using Exponential Functions

1. Model the exponential growth situation of your biologist friend's paramecia, noting that each hour the population doubles.

Suppose t is the number of hours that have passed and $A(t)$ is the number of paramecia at the end of t hours. If she begins her experiment with one paramecium, then $A(0) = 1$. Moreover, since the population doubles by the end of each hour, $A(1) = 2$, $A(2) = 4$, and so on.

a. Calculate the results of the experiment at the end of each of the first 5 hours by completing the following table.

t	0	1	2	3	4	5
$A(t)$	1	2				

b. Since the value of $A(t)$ can be expressed as a power of 2, a mathematical description of your friend's situation is

$$A(t) = 2^t, \text{ where } t \geq 0$$

Use this function to calculate the number of paramecia at some particular times and also to determine the time when a given number exists.

(1) Since the population is continually increasing, it makes sense not only to talk about the number of paramecia at the end of each hour but also at each moment in time. Use ISETL, your CAS, or your calculator to approximate the number of paramecia at the end of the following times. Be as accurate as possible. Round your answers down to the nearest whole paramecium.

(a) Find $A(t)$ when $t = 8$—that is, find the number of paramecia after 8 hours.

(b) Find $A(t)$ at the end of 10 hours, 15 minutes.

(c) Find $A(t)$ when $t = 17$.

(d) Find $A(t)$ after 20 hours, 30 minutes.

(e) Find $A(t)$ at the end of 2 days.

(f) Find $A(t)$ after 100 hours.

3

(2) Approximate as accurately as you can the number of hours which need to pass before your friend has the following number of paramecia:

(a) 1,000 paramecia

(b) 100,000 paramecia

(c) 1 million paramecia

c. After a short period of time, the function A grows very rapidly. Use your CAS to examine the graph of A. Place the graphs for the specified values of t in your Activity Guide.

(1) $0 \leq t \leq 5$

(2) $0 \leq t \leq 10$

2. Consider a different initial condition. Examine how this change affects the expression representing the situation and how it affects the graph.

a. Suppose she starts her experiment with 10 paramecia (instead of 1). Model the new situation with a function called *S*.

(1) Calculate the number of paramecia at the end of each hour, for the first 5 hours. Express the result in terms of a power of 2. Note the impact of the asumption that initially she has 10 paramecia.

t (hrs)	$S(t)$ (value)	$S(t)$ (in terms of a power of 2)
0	10	$10 \cdot 2^0$
1	$2 \cdot 10$ or 20	$10 \cdot 2^1$
2		
3		
4		
5		

(2) Represent *S* by an expression.

$S(t) =$, where $t \geq 0$

(3) Since the population is doubling, the expression representing *A* (which you considered in part 1) and the one representing *S* both have the same base, namely 2. The only thing that differs is their initial condition, since by assumption $A(0) = 1$ whereas $S(0) = 10$.

Graph *A* and *S* on the same pair of axes for different values of *t*; for example, for $0 \leq t \leq 2$ and for $0 \leq t \leq 5$. Describe the

3

relationship between the two graphs. Support your description with a sketch.

3. Consider another scenario. Suppose something goes wrong and your friend's paramecia start to die. When she initially notices the problem she has a million paramecia. One hour later she has half that many, and after each subsequent hour she has half again as much. Model the new situation with an exponential decay function called *D*.

a. Calculate some values of $D(t)$.

t (hrs)	$D(t)$ (value)	$D(t)$ (in terms of a power of $\frac{1}{2}$)
0	1,000,000	$1{,}000{,}000 \cdot (\frac{1}{2})^0$
1	$\frac{1}{2} \cdot 1{,}000{,}000$ or 500,000	$1{,}000{,}000 \cdot (\frac{1}{2})^1$
2		
3		
4		
5		

b. Represent *D* by an expression.

$D(t) =$ where $t \geq 0$

c. Answer some questions about the situation.

(1) Find the population at the end of 4 hours, 30 minutes.

(2) During what 15-minute time interval does the population drop below 20,000?

d. Describe the shape and location of the graph of *D*. Support your description with a sketch.

In the last task you used exponential functions to model several situations. You analyzed two growth functions, namely $A(t) = 2^t$ and $S(t) = 10 \cdot 2^t$, and a decay function, $D(t) = 1{,}000{,}000 \cdot (\frac{1}{2})^t$. You compared the shapes of their graphs and observed how changing the initial condition affects the location of the graph. The domains of the exponential functions which you considered do not contain any negative numbers, since in your friend's experiment it doesn't make sense to talk about negative time. Exponential functions, however, can be defined for all real numbers.

In general, if a is a fixed positive number, not equal to 1, then

$$f(x) = c \cdot a^x, \quad \text{where } -\infty < x < +\infty$$

is called an *exponential function* with base a and initial condition c. When $0 < a < 1$, f is an *exponential decay* function. When $a > 1$, f is an *exponential growth* function. Since $f(0) = c$, the value of c determines where the graph crosses the vertical axis.

In the next task, you will compare the behavior of exponential growth functions and the behavior of exponential decay functions for different values of the base. You will also compare the behavior of an exponential growth function, such as $g(x) = 3^x$, and its associated exponential decay function, $d(x) = (\frac{1}{3})^x$.

Before beginning Task 3-10, recall two rules for manipulating exponents:

$$\left(\frac{1}{a}\right)^x = \frac{1}{a^x} = a^{-x}$$

and

$$a^{m/n} = \sqrt[n]{a^m}$$

According to these rules, $(\frac{1}{8})^x$ and $\frac{1}{8^x}$ and 8^{-x} are three different ways of writing the same thing, and

$$8^{-2/3} = \frac{1}{8^{2/3}} = \frac{1}{\sqrt[3]{8^2}} = \frac{1}{\sqrt[3]{64}} = \frac{1}{4}$$

Task 3-10: Comparing Exponential Functions

1. Analyze how the value of the base affects the shape and location of an exponential growth function.

a. Consider the behavior of three exponential growth functions whose bases are $a = 2$, $a = 3$, and $a = 5$. Let

$$f(x) = 2^x$$
$$g(x) = 3^x$$
$$h(x) = 5^x$$

(1) Compare some output values for f, g, and h.

Evaluate the functions at a variety of inputs, both negative and positive. Record your results in the table given below.

x	$f(x) = 2^x$	$g(x) = 3^x$	$h(x) = 5^x$

(2) Compare the graphs of f, g, and h.

Use your CAS to plot f, g, and h on same pair of axes, for $-1 \leq x \leq 5$. To get a better idea about what's happening, plot the graphs for $-1 \leq x \leq 1$.

Print a copy of the graphs and paste it in below. Label each function.

(3) Summarize your observations concerning the behavior of f, g, and h.

Compare the values of the three functions and describe the relationship among the graphs, when x is negative, when x equals zero, and when x is positive.

b. Generalize your observations concerning the impact of the size of the base on the shape and location of the graph of an exponential growth function.

Let g and h be two arbitrary functions, having different bases, where

$$g(x) = p^x \quad \text{and} \quad h(x) = r^x, \quad \text{where } 1 < p < r$$

Note: All you know about the bases p and r is that they are two real numbers, they are both greater than 1, and p is less than r. You do not know their actual values.

(1) Sketch (by hand) representative graphs of g and h on one pair of axes. Label the two graphs.

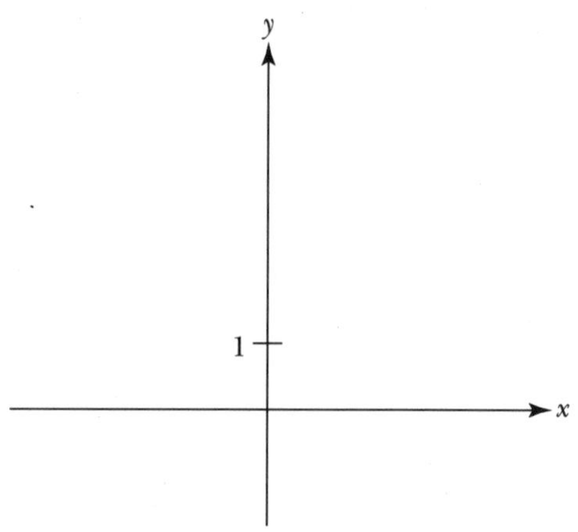

(2) Describe the set of all x values such that $g(x) < h(x)$—that is, the set of all values where the graph of g lies below the graph of h.

(3) Describe the set of all x values such that $g(x) > h(x)$—that is, the set of all values where the graph of g lies above the graph of h.

(4) Describe the set of all x values such that $g(x) = h(x)$—that is, the set of all values where the two graphs intersect.

2. Investigate how the value of the base affects the shape and location of an exponential decay function.

a. First, use your CAS to scrutinize some particular decay functions and note their relationships.

(1) Consider $g(x) = (\frac{1}{4})^x$ and $h(x) = (\frac{1}{2})^x$. Compare their values when x is negative, when x equals zero, and when x is positive. Compare the shape and location of the two graphs. Sketch the graphs of the functions on the axes given below. Label the graphs, indicating the value of the base for each function.

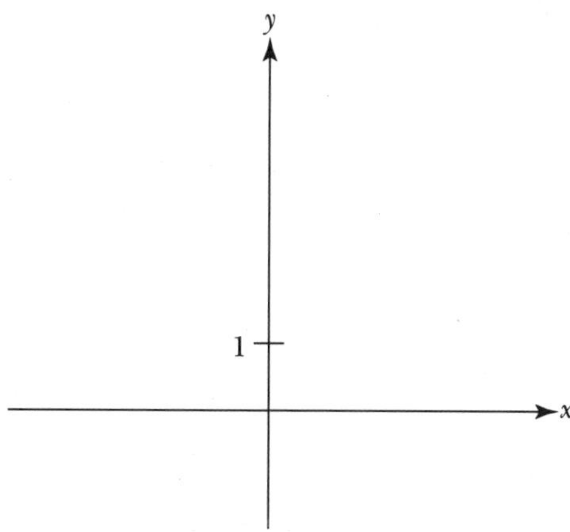

(2) Repeat the process for another pair of values for the base. For instance, compare the decay functions generated by $a = \frac{1}{3}$ and $a = \frac{2}{3}$. Sketch their graphs on the axes given below. Label the graphs.

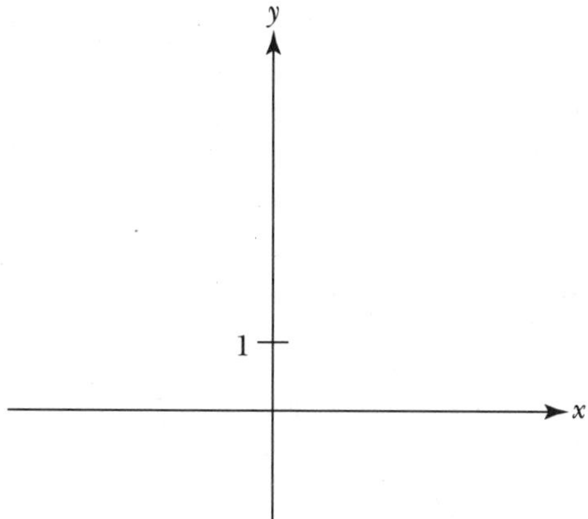

b. Generalize your observations from part a.

(1) On the pair of axes given below, sketch graphs of two arbitrary exponential decay functions g and h, where $g(x) = u^x$, $h(x) = w^x$, and $0 < u < w < 1$. Label the graphs.

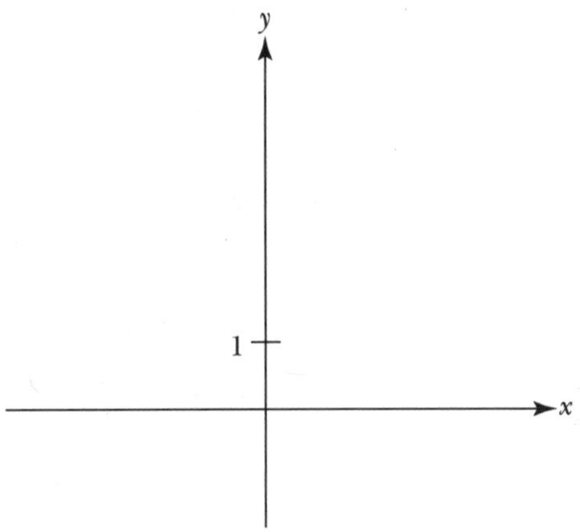

(2) Describe the set of all x values such that $g(x) < h(x)$.

(3) Describe the set of all x values such that $g(x) > h(x)$.

(4) Describe the set of all x values such that $g(x) = h(x)$.

3. Compare a growth function to its associated decay function.

Notice that if the value of the base is greater than 1—that is, if $a > 1$—then the reciprocal of the base is a positive number less than 1—that is, $0 < 1/a < 1$. Consequently, for each value of $a > 1$, there exists an exponential growth function,

$$g(x) = a^x, \quad \text{where } -\infty < x < +\infty$$

and a corresponding exponential decay function

$$d(x) = \left(\frac{1}{a}\right)^x = \frac{1}{a^x} = a^{-x}, \quad \text{where } -\infty < x < +\infty$$

Find the relationship between the graphs of these associated functions.

a. For each of the following exponential functions, state whether it is a growth or decay function. If it is a growth function, give the expression for its associated decay function. If it is a decay function, give the expression for its associated growth function.

(1) $f_1(x) = 3^x$

(2) $f_2(x) = \dfrac{1}{8^x}$

(3) $f_3(x) = 4^{-x}$

(4) $f_4(x) = 0.25^x$

(5) $f_5(x) = 2.5^x$

(6) $f_6(x) = \left(\dfrac{1}{5}\right)^x$

(7) $f_7(x) = \left(\dfrac{11}{2}\right)^x$

b. Compare the graphs of a growth function and its corresponding decay function.

(1) Consider some particular functions.

(a) For instance, let $a = 2$. Use your CAS to graph $g(x) = 2^x$ and $h(x) = (\frac{1}{2})^x$ on the same pair of axes. Describe the relationship between the shapes and locations of the two graphs.

3

Sketch the graphs of the associated functions below. Label the functions.

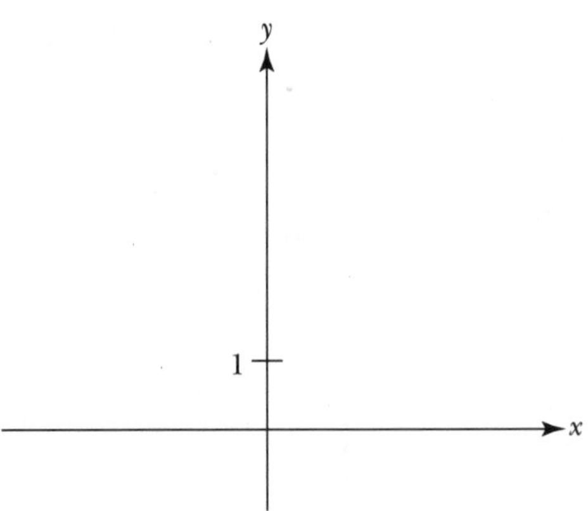

(b) Repeat the process for another pair of functions. For instance, consider $a = 4$. What do you notice about the shape and location of the graphs for this new value? Sketch the graphs of this pair of associated functions on the axes given below. Label the graphs.

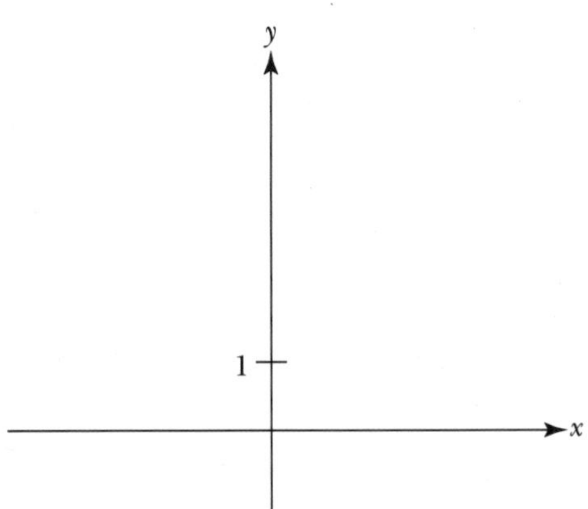

(2) Suppose $a > 1$. Describe the relationship between the graphs of an arbitrary growth function, $g(x) = a^x$, and its associated decay function, $d(x) = (1/a)^x$.

c. Compare the expressions for a growth function and its corresponding decay function.

Recall that to represent the reflection of a function through the y-axis by an expression, you replace each x in the given expression with $-x$. Use this fact to show that $g(x) = a^x$ and $d(x) = (1/a)^x$ are reflections of one another through the vertical axis.

4. Summarize your observations.

$f(x) = a^x$	$0 < a < 1$	$1 < a$
Sketch of typical graph		
Domain		
Range		
x-intercept(s)		
y-intercept		
Intervals where increasing		
Intervals where decreasing		
Intervals where concave up		
Intervals where concave down		
Intervals where one-to-one		

In the last task, you examined the behavior of exponential growth functions and exponential decay functions. You investigated the general shape of the graph of a growth function and the shape of a decay function; you compared the graphs of growth and decay functions that have different bases; you analyzed the relationship between a growth function and its corresponding decay function and observed that the two functions are reflections of one another through the vertical axis.

Now turn your attention to the class of functions called *logarithmic functions*, or *log functions* for short. Just as exponential growth and decay func-

3

tions are reflections of one another, logarithmic functions and exponential functions are also reflections. In particular, they are inverse functions.

The term "inverse function" should sound familiar. In Unit 2, Section Three you learned how to develop new functions from existing ones by performing reflections. In particular, given a one-to-one function f, you constructed its *inverse function* f^{-1} by reflecting f through the line $y = x$. Finding the reflection of a graph through $y = x$ involved flipping the graph over the diagonal line. Representing the reflection of a function through $y = x$ by an expression involved interchanging the x's and y's in the expression for the given function and solving for y.

Since an exponential function is a one-to-one function, it makes sense to find its inverse. This is the approach which you will use in the next task to develop the concept of a logarithmic function.

Task 3-11: Investigating the Relationship Between Exponential and Logarithmic Functions

1. Examine the reflection of a given exponential function through the line $y = x$.

Consider the exponential function with base 2, $y = 2^x$, where $-\infty < x < +\infty$.

a. Find some values of $y = 2^x$ by filling in the following table.

x	−2	−1	0	1	2
y					

b. Graph $y = 2^x$ and its reflection through $y = x$ on the same pair of axes. Use the entries in the table for $y = 2^x$, recalling that you can graph the reflection by graphing the ordered pairs in reverse order.

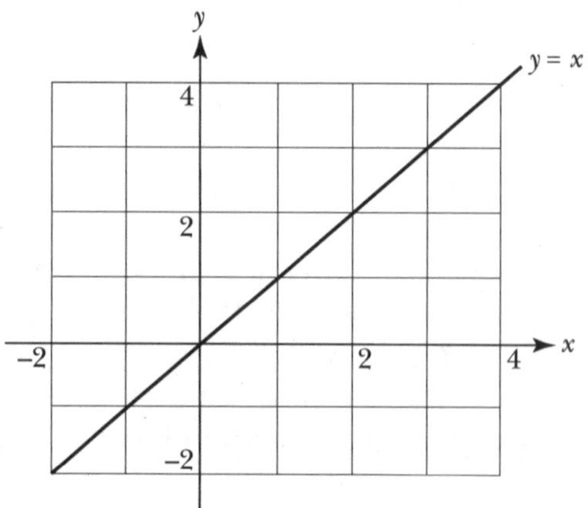

c. The reflection of $y = 2^x$ through the line $y = x$ is called a *logarithmic function with base 2.* Try to represent the reflection by an expression.

(1) Interchange the x's and y's in $y = 2^x$.

(2) *Try* to solve for y.

Observation: After interchanging the x's and y's in the given exponential expression, $y = 2^x$, it is impossible to solve $x = 2^y$ for y algebraically. Instead, you write

$$y = \log_2(x)$$

which is read, "y equals log base 2 of x." The important thing to realize is that $x = 2^y$ and $y = \log_2(x)$ are simply two different ways of referring to the same function; the first is called the exponential form of the logarithmic function base 2 and the second is called the logarithmic form.

2. Graph some given exponential functions and their corresponding logarithmic functions on the same pair of axes.

First, let's generalize the comments about logarithmic notation which were made above. If a is a positive real number, not equal to 1, you can find the inverse of exponential function $y = a^x$ as follows:

$y = a^x$	Given
$x = a^y$	Interchange x and y
??	Solve for y

Since there is no straightforward way of solving $x = a^y$ for y, the notation

$$y = \log_a x$$

which is read, "y equals log base a of x," is used to denote the inverse of $y = a^x$. It is called the *logarithmic* function with base a. In other words:

- $y = \log_a x$ is equivalent to $x = a^y$.
- $y = \log_a x$ is the reflection of $y = a^x$ through the line $y = x$.

Now, consider some specific functions.

Note: The following exponential functions are ones you considered in the last task.

3

a. Find the logarithm associated with the exponential growth function $y = 3^x$.

(1) Give the exponential form of the associated logarithm.

(2) Give the logarithmic form of the associated logarithm.

(3) Plot a few points on the graph of $y = 3^x$. Use this information to sketch a graph of $y = 3^x$ and the graph of its associated logarithm. Label the two graphs. Check your graphs using your CAS.

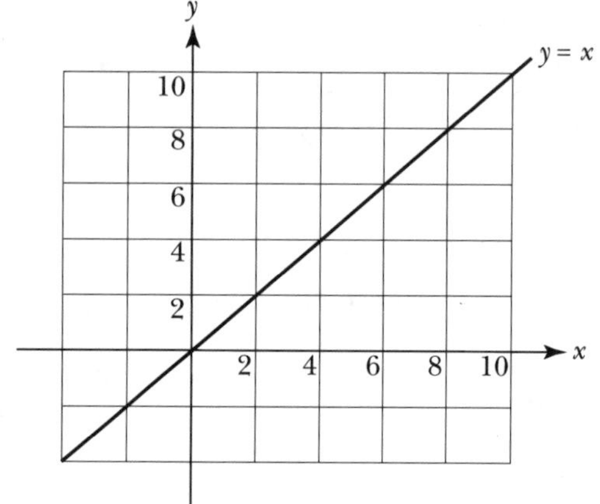

(4) Recall that the domain of $y = 3^x$ is the set of all real numbers, or $-\infty < x < +\infty$. Give the domain of its associated logarithmic function.

b. Find the logarithm associated with the exponential decay function $y = (\frac{1}{2})^x$.

(1) Give the exponential form of the associated logarithm.

(2) Give the logarithmic form of the associated logarithm.

(3) Plot a few points on the graph of $y = (\frac{1}{2})^x$. Use this information to sketch graph of $y = (\frac{1}{2})^x$ and the graph of its associated logarithm. Label the two graphs. Check your answer using your CAS.

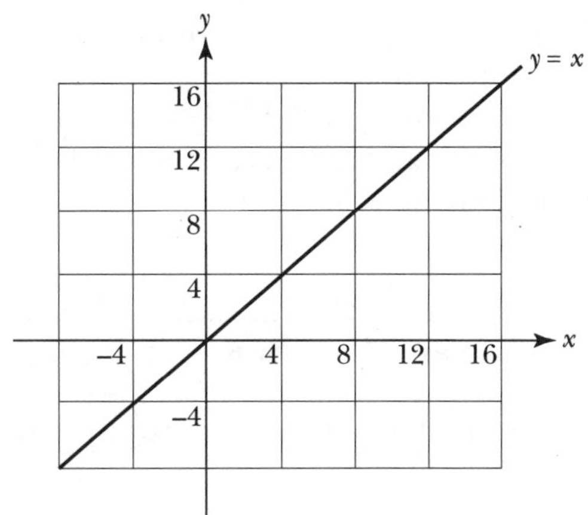

(4) Give the domain of the associated logarithmic function.

3. Generalize your observations concerning logarithmic functions.

$y = \log_a x$	$0 < a < 1$	$1 < a$
Sketch of a typical graph		
Domain		
Range		
x-intercept(s)		
y-intercept		
Intervals where increasing		
Intervals where decreasing		
Intervals where concave up		
Intervals where concave down		
Intervals where one-to-one		

3

In the last task, you examined the relationship between a given exponential function and its associated logarithmic function.

The next task will give you some practice working directly with logarithmic functions. In particular, you will:

- Convert back and forth between the two representations for logs, namely the exponential and logarithmic forms.
- Examine ways to evaluate a logarithmic expression, such as $\log_2 128$.
- Examine ways to graph a logarithmic function directly, without first graphing its associated exponential function.

Task 3-12: Evaluating and Graphing Log Functions

1. Practice using logarithmic notation by converting between exponential and logarithmic forms.

As stated above, $5^y = x$ is equivalent to $y = \log_5 x$, where $5^y = x$ is in exponential form, and $y = \log_5 x$ is in logarithmic form. The following table gives some equations in exponential and/or logarithmic forms. Fill in the missing entries.

Exponential form	Logarithmic form
$2^4 = 16$	$\log_2 16 = 4$
$5^3 = 125$	
	$\log_5 1 = 0$
	$\log_7 7 = 1$
$\pi^2 = 9.86 \ldots$	$\log_\pi 9.86 \ldots = 2$
$2^{-3} = (\frac{1}{8})$	
$e = 2.718 \ldots$	$\log_e 2.718 \ldots = 1$
$10^4 = 10{,}000$	
	$\log_6(\frac{1}{6}) = -1$
$\pi^1 = 3.2415 \ldots$	
	$\log_2 \sqrt[3]{2} = \frac{1}{3}$

2. Evaluate some logarithmic expressions.

a. One way to evaluate a logarithmic expression is to write the expression as an equation, convert the equation to exponential form, and then equate the exponents.

For example, to evaluate $\log_{10} 100$ you would do the following:

$\log_{10} 100$	Given
$\log_{10} 100 = y$	Write as an equation
$100 = 10^y$	Convert to exponential form
$10^2 = 10^y$	Express left side as a power of the base, namely 10
$y = 2$	Equate exponents

Evaluate the following logs using this method. Use your CAS or your calculator to check your answers.

(1) $\log_{10} 1$

(2) $\log_{10} 0.1$

(3) $\log_{10} 0.001$

(4) $\log_{10} 10^6$

(5) $\log_2 \sqrt[3]{2}$

(6) $\log_2 0.5$

(7) $\log_{0.5} 1$

(8) $\log_{0.5}\left(\frac{1}{2}\right)^{-7}$

3

(9) $\log_{0.5}\left(\frac{1}{16}\right)$

b. You can also evaluate logs directly without writing an equation and converting its form. The key is knowing how to "read" logs.

For example, in order to evaluate $\log_{10} 100$, you could ask yourself:

Question: What power do I raise 10 to in order to equal 100?
Answer: Since 10 raised to the power 2 equals 100, the answer to the question is 2.
Therefore, $\log_{10} 100 = 2$.

Evaluate the following logs by translating the expression to the related question and then finding the answer. Use your CAS or your calculator to check your answers.

(1) Evaluate $\log_2 32$.

Question:

Answer:

Therefore, $\log_2 32 =$

(2) Evaluate $\log_3 81$.

Question:

Answer:

Therefore, $\log_3 81 =$

(3) Evaluate $\log_4 64$.

Question:

Answer:

Therefore, $\log_4 64 =$

3. Graph some logarithmic functions directly (without sketching the associated exponential function).

A couple of comments before you begin:

i. It is important to recognize that $y = \log_a x$ and $y = a^x$ are inverse functions, whereas $y = \log_a x$ and $a^y = x$ are equivalent forms of the same equation.

ii. When graphing $y = \log_a x$, remember that x is the independent variable and y the dependent variable.

a. Graph a familiar log function using the usual approach.

The "usual way" to graph a function is to choose values for the independent variable x, solve for the corresponding values of the dependent variable y, and then plot the points. Graph $y = \log_2 x$ using the usual approach.

(1) It is to your advantage to choose the x values "wisely," such as $x = 1, 2, 4, \frac{1}{2}, \frac{1}{4}, \frac{1}{8}$. Wisely choose three more values for x and record the corresponding y values in the table.

x	$\frac{1}{8}$	$\frac{1}{4}$	$\frac{1}{2}$	1	2	4			
y									

(2) Explain what it means to make a "wise" choice.

(3) Graph the function $y = \log_2 x$ by plotting the points in your table. Label the axes.

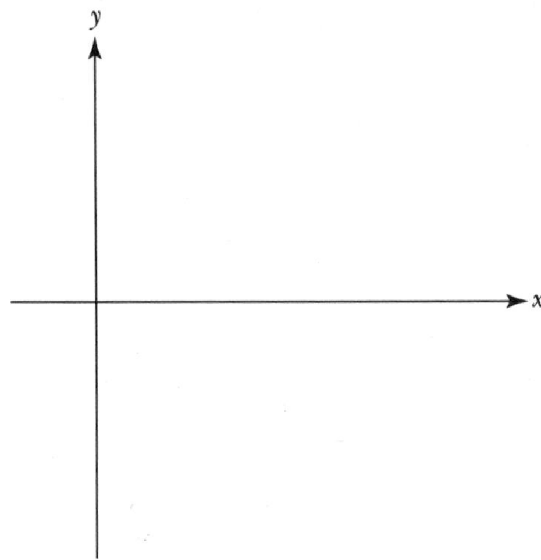

3

b. Graph a familiar log function by considering the exponential form.

Another way to graph a log function is to convert the equation $y = \log_a x$ to exponential form $a^y = x$. Then, instead of choosing values for the independent variable x, you choose values for the dependent variable y and solve for x. In other words, you reverse the process used in the usual approach. Graph $y = \log_2 x$ using this approach.

(1) Convert $y = \log_2 x$ into exponential form.

(2) Find ordered pairs that satisfy the equation $2^y = x$ by choosing values for y and then solving for x.

Build a table of 10 ordered pairs that satisfy $2^y = x$. Note: Some values for y are already entered.

x										
y	−3	−2	−1	0	1	2	3			

(3) Graph the function $y = \log_2 x$, by plotting the points in your table. Label the axes.

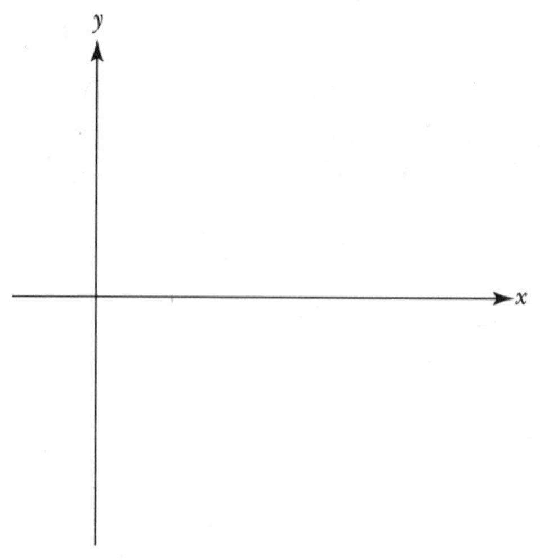

c. Use your CAS to check that your graphs of $y = \log_2 x$ are correct. Plot the function with x varying from $\frac{1}{8}$ to 8. Place the graph in the space below.

d. Examine the domain of a logarithmic function.

The domain of $y = \log_2 x$ appears to be $x > 0$. Investigate this a little more closely. Rewrite $y = \log_2 x$ in its exponential form $2^y = x$. Can you find any value for y that is associated with a negative value of x? Can you find a value for y that is associated with $x = 0$? Justify your responses.

Unit 3 Homework After Section 3

- Complete the tasks in Section Three in the Activity Guide. Be prepared to discuss them in class.
- Convert between forms, evaluate some logarithms, and sketch and analyze graphs of some related functions in HW3.11.

HW3.11 Practice.

1. Convert the following equations to logarithmic form.

a. $6^3 = 216$

b. $(\frac{1}{3})^5 = \frac{1}{243}$

c. $7^p = q$

d. $a^b = c$

e. $2.5^{-2} = 0.16$

f. $e^2 \approx 7.389$ *Note:* $e = 2.718281828...$

2. Convert the following equations to exponential form.

a. $\log_4 64 = 3$

b. $\log_{\frac{1}{8}} 4096 = -4$

c. $\log_2 \frac{1}{8} = -3$

d. $\log_{10} 10^5 = 5$

e. $\log_m n = p$

f. $\ln(20.085) \approx 3$ *Note:* $\ln(x) = \log_e x$

3. Evaluate the following logs.

a. $\log_2 64$

b. $\log_3 \frac{1}{9}$

c. $\log_{2.5} 1$

d. $\log_5 \sqrt[3]{5^2}$

e. $\log_{0.5} (\frac{1}{2})^{-12}$

f. $\log_{\frac{1}{4}} 16$

4. For each of the following (related) functions:

i. Sketch the general shape of the function.

ii. Give the domain and range of the function.

iii. Describe the set of all x values where the function is increasing or decreasing.

iv. Describe the set of all x values where the function is concave up or down.

v. Give the set of all turning points.

a. $f(x) = 4^x$

b. $f(x) = \dfrac{1}{4^x}$

c. $f(x) = \log_4 x$

d. $f(x) = \log_{\frac{1}{4}} x$

- Reflect some exponential and logarithmic functions through various lines in HW3.12.

HW3.12 For each of the three functions given below in 1–3, reflect the function

a. through the y-axis.

b. through the x-axis.

c. through the line $y = x$.

For *each* reflection of the given function:

(1) Represent the specified reflection by an expression.

(2) Sketch the general shape of the function and the specified reflection on a single pair of axes. Label the two graphs.

(3) Give the domain of the reflection.

1. $y = e^x$, where x is any real number.

Note: As you know, the base of an exponential or logarithmic function can be any positive number not equal to 1. A special base, which you will examine more closely later on, is the number e, where e = 2.718281828... The logarithm base e, $\log_e x$, is called the natural logarithm function; it is also denoted by ln(x).

2. $y = \log_{10} x$, where $x > 0$.

3. $y = \dfrac{1}{3^x}$, where x is any real number.

- Describe how the various exponential and logarithmic functions are related in HW3.13.

HW3.13 Suppose a is an arbitrary positive real number, not equal to 1.

1. Describe the relationship between $y = a^x$ and $y = a^{-x}$.

2. Describe the relationship between $y = a^x$ and $y = \log_a x$.

3. Describe the relationship between $x = a^y$ and $y = \log_a x$.

- Consider some properties of logarithmic in HW3.14.

HW3.14 Properties of logarithms.

1. Prove the following properties, where a is a positive real number not equal to 1, by converting the given equation to exponential form.

a. $\log_a 1 = 0$

b. $\log_a a = 1$

c. $\log_a a^x = x$, for any real number x

2. Suppose a, r and s are positive real numbers, where a is not equal to 1. Some additional properties of logarithms are

i. $a^{\log_a r} = r$

ii. $\log_a rs = \log_a r + \log_a s$

iii. $\log_a \dfrac{r}{s} = \log_a r - \log_a s$

iv. $\log_a r^c = c \log_a r$, for every real number c

Note: These properties can be proved using the laws of exponents and the fact that logarithmic and exponential functions are inverse functions.

3

a. Use the properties of logs to express each of the following as a combination of $\log_a x$, $\log_a y$, and $\log_a z$.

(1) $\log_a \frac{(x^2y)^2}{z}$ (3) $\log_a \frac{z\sqrt{x}}{(xy)^{\frac{1}{4}}}$

(2) $\log_a \sqrt{\left(\frac{x}{y}\right)^3}$ (4) $a^{\log_a(\log_a xy)}$

b. Use the properties of logs to express each of the following as the log of a single expression.

(1) $4 \log_a x - 2 \log_a y - 3 \log_a z$ (3) $6 \log_a(a^{\log} ax)$

(2) $\frac{1}{4} \log_a z + \frac{1}{2} \log_a y + \frac{3}{4} \log_a x$ (4) $\log_a(a^{2 \log} ax) - \log_a(a^{3 \log} ay)$

- Compare the growth of an exponential function to a power function in HW3.15.

HW3.15 Consider the following salary negotiations between a king and a wanderer.

King's offer: I'll pay you 1¢ for the first day of work, and then the square of the number of the day for each day thereafter. That is, I'll pay you 1¢ for day 1, 4¢ for day 2, 9¢ for day 3, and so on.

Wanderer's request: Please pay me 2¢ for the first day and then twice as much (as the previous day's pay) for each day thereafter. In other words, pay me 2¢ for day 1, 4¢ for day 2, 8¢ for day 3, and so on.

Since the king was desperate for help, and the wanderer's request seemed reasonable, the king agreed.

1. Represent the king's offer by a function K, where $K(t)$ is the amount of money (in cents) the king is willing to pay the wanderer for day t, where $t = 1, 2, 3, 4, \ldots$

2. Represent the wanderer's request by a function W, where $W(t)$ is the amount of money (in cents) the wanderer wants to be paid for day t, where $t = 1, 2, 3, 4, \ldots$

3. Consider the wanderer's request.

a. On which day would the king pay the wanderer $20.48?

b. What is the first day on the king would pay the wanderer more than $10,000?

c. How much would the king pay the wanderer on the 30th day?

4. Compare the king's offer and the wanderer's request.

a. On which day(s) would the wanderer receive the same amount with either plan?

b. Based on the two plans, what is the difference in the amount the wanderer would be paid on:

(1) day 10?
(2) day 15?
(3) day 20?

c. Graph the functions modeling the king's offer and the wanderer's request on the same pair of axes. Compare the growth rates of the two functions as the days go by.

d. Did the king make the right decision?

SECTION 4

Fitting Curves to Discrete Functions

Function classes enable you to fit a curve to—or find a model of—a set of real data. You can then use the model to analyze the data and make predictions. For example, suppose you are conducting a study of the number of HMOs (health maintenance organizations) in the United States between 1978 and 1990, and gather the following data, where $t = 0$ corresponds to the baseline for your study, 1978, when there were 5,000 HMOs.

t	0	2	4	6	8	10	12
Year	'78	'80	'84	'84	'86	'88	'90
# HMOs (100's)	50	1203	1499	2089	2306	2356	2650

Table 1.

The graph of this discrete data is represented by the scatter plot given below. Rather than working directly with a bunch of unconnected points, it is often much easier to try and find a continuous curve that "fits" or models the data—that is, an expression whose graph is very similar to the scatter plot representing the given data. In this instance, the data have a logarithmic shape, and the function $f(t) = 10^3 \ln(t + 1) + 50$, whose graph is shown below by the curved line, appears to fit the data fairly closely.

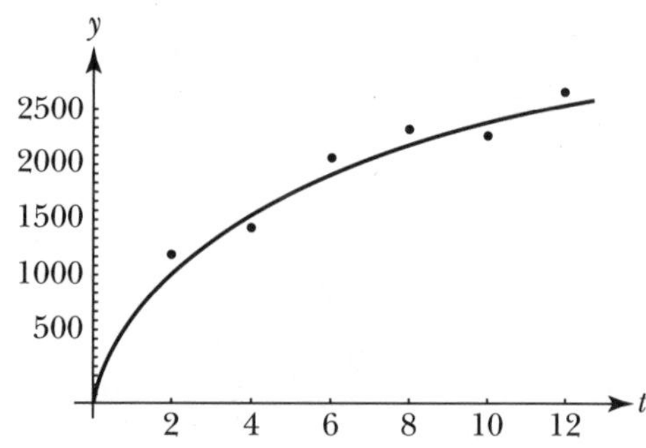

Graph of Data in Table 1 and $f(t) = 10^3 \ln(t + 1) + 50$

The new fit function can be used to predict what happens at points that are not in the original data set. For example, to predict how many HMOs there were in 1981, you could calculate $f(3)$. Be careful, however. If you use the fit function to predict what happens at values that lie above or below the endpoints of the original domain, scrutinize your results with care. The fit function may behave very differently outside the domain of the data set and thus give inaccurate approximations.

Of course, not all data can be fitted nicely with one curve. Sometimes you have to use several curves and model the data with a piecewise-defined function.

In the next task, you will use your graphing package to fit curves to some discrete functions and make predictions about the data for values not in the domain of the data set. Since finding a curve which best fits the data typically involves plotting the data and guessing the general shape of its graph, you will start by reviewing the general shape of some fundamental function classes.

Task 3-13: Modeling Data

1. First, recall the general shape of some basic functions.

Draw a rough sketch indicating the general shape of the following functions.

a. $f(x) = mx + b$, where $m \neq 0$

b. $f(x) = ax^2 + bx + c$, where $a \neq 0$

c. $f(x) = a_3x^3 + a_2x^2 + a_1x + a_0$, where $a_3 \neq 0$

d. $f(x) = \ln(x)$, where $x > 0$

e. $f(x) = e^x$

f. $f(x) = \sin(x)$

2. Review the section in the course handout for your CAS that describes how to use your CAS to fit a curve to a given data set.

3. Model some situations with a function.

For each of the following discrete functions:

i. Represent the discrete function by a scatter plot.

ii. Find a continuous curve that fits the data as closely as possible. Record your function in your Activity Guide.

iii. Plot the discrete function and the curve fit on the same pair of axes. Print a copy of the graph and place it in your Activity Guide.

iv. Use the fit function to answer any questions pertaining to the data set.

a. Make your first pass at fitting a curve to some data while thinking about a yak named Charlie.

A mountain yak named Charlie is sprinting across a field moving at a nearly constant rate. By measuring Charlie's distance at 1-second intervals from a fixed point you have collected the following data:

Time (s)	Distance (m)
0	0.0
1	20.4
2	39.8
3	60.0
4	79.5
5	99.8
6	120.0
7	143.1
8	159.7
9	181.0
10	200.5

(1) Represent Charlie's movement with a scatter plot.

(2) Describe the shape of the scatter plot. What type of function might you use to fit the data?

(3) Since Charlie is moving at nearly a constant rate, his distance versus time graph is almost linear. Fit a linear function to the data. Record your function below.

(4) Graph the original data and the function modeling Charlie's movements on the same pair of axes. Print a copy of the graph and place it in below.

(5) Use your fit function to estimate Charlie's distance at:

(a) 2.5 seconds

(b) 7.9 seconds

(c) 0.25 seconds

(6) Find the size of the error—that is, the difference between the actual data and the value of the fit function—at 7 seconds.

b. Fit a curve to data collected from moving in front of the motion detector.

Suppose for 13 seconds you walked back and forth in front of the motion detector and collected the following data:

Time (s)	0	1	2	3	4	5	6	7	8	9	10	11	12	13
Distance (m)	2	3.26	3.36	2.21	0.86	0.57	1.58	2.99	3.48	2.62	1.18	0.5	1.19	2

(1) Plot the data that you collected on a scatter plot.

(2) Find a curve that fits your distance versus time graph. Record your function below.

(3) Plot the data that you collected and the curve that models your movement on the same pair of axes. Make a printout of the graph and place it in below.

(4) Using the function that models your movement, give a detailed description of the way you moved in front of the detector. Include in your description approximations of

- the time intervals when you were increasing your distance from the detector.
- the time intervals when you were decreasing your distance from the detector.
- the times when you reversed direction.

c. Fit a function to a population growth situation.

Your scientist friend is at it again, growing little beasties in Petri dishes in her lab. At first it is fairly easy to keep track of how many there are, but then the total number starts increasing faster than she can count. Before she admits defeat, however, she manages to collect the following data:

Time (min.)	0	0.5	1.0	1.5	2.0	2.5	3.0	3.5	4.0
Number of beasties	6	10	15	27	42	75	125	200	325

(1) Plot the data that your friend has collected.

(2) Fit the data with a continuous function. Record the function in the space below.

(3) Graph the data and the curve on the same pair of axes. Paste a copy of the graph in your Activity Guide.

(4) Use the function that fits the data to estimate the number of beasties that your friend had at the following times. (Since fractional beasties don't make much sense, round your responses down.)

(a) At 20 seconds.

(b) At one minute and forty-five seconds.

(c) At 3 minutes and 15 seconds.

d. Model total sales in your pizza business.

Five years ago you bought a pizza place in the town of Collegeville. During the past year business has been phenomenal—both home deliveries and your take-out service have increased rapidly. Unfortunately, times haven't always been as good, as the following sale records show:

Year (t)	Total sales ($1000's)
0.5	24
1.0	33
1.5	37
2.0	38
2.5	36
3.0	37
3.5	38
4.0	42
4.5	51
5.0	65

(1) Noting that when you first started out—that is, when t equaled 0—you had no sales, find a curve that models your sales data during the five years that you've been in business. Name the model S. Record the function below.

(2) Graph your data and S on the same pair of axes. Print a copy of the graph and paste it below.

(3) Use S to predict your total sales at the following times. Show how you arrived at your answer.

(a) Two years and nine months after you opened.

(b) Four months ago.

(c) At the end of the first two months.

(d) Thirty-nine months after you opened.

(4) If you were to use S to predict your total sales 5 years from now, what would your estimate be? How reliable is this estimate?

(5) Based on your model, describe the way your business has grown during the past 5 years. During what time periods were sales increasing? Decreasing? During which 6-month period did you have your fastest growth?

(6) Analyze the accuracy of your model at some specified times by filling in the following chart:

Year (t)	Total sales ($1000's)	$S(t)$	Error
0			
2.0			
2.5			
3.0			
3.5			
5.0			

Unit 3 Homework After Section 4

- Complete the tasks in Section Four in the Activity Guide. Be prepared to discuss them in class.

HW3.16 Analyze two supply and demand functions.

Supply Data

Price ($)	Number of tickets
0	0
20	4,000
40	13,000
60	20,000
80	36,000
100	50,000

You have agreed to be the promoter of a major outdoor concert. There are an incredible number of little details to which you must attend, but one of the major decisions you must make is how much to charge for each ticket. You realize that the number of tickets you would be willing to sell increases as the price of the tickets increases, since your net profit will be greater and your ticket selling operation will be operating at increasingly efficient levels. After some back-of-the-envelope calculations you come up with the data (given in the table on the left) concerning the number of tickets you would be willing to supply at various prices.

Although you're willing to supply more tickets as the price increases, the potential ticket buyers feel exactly the opposite way. The number of tickets they are willing to buy goes

Demand Data

Price ($)	Number of tickets
15	50,000
35	30,000
55	10,000
75	5,000
95	2,500

down as the price per ticket increases. To try to determine the number of tickets that will be demanded by the customers at various prices, you hire a marketing firm to take a survey. The table given to the left summarizes their report.

1. Find a function that models your supply data. Name your function *S*.
2. Find a function *D* that models the consumers' demand data. Name your function *D*.
3. Graph the four functions—the supply data, the demand data, and the two fit functions, *S* and *D*—on the same pair of axes. Print a copy of the graph. Label the four graphs.
4. Use *S* and *D* to approximate solutions to the following questions. Show your work.
   - **a.** If the tickets cost $25 each, how many tickets would the public be willing to buy?
   - **b.** If the tickets cost $75, how many tickets would you be willing to sell?
   - **c.** If the tickets cost $65, what is the difference between the number of tickets you would be willing to promote and the number the public would be willing to buy?
   - **d.** An equilibrium point is the point where the supply and demand graphs intersect.
     - (1) Use *S* and *D* to find the price of tickets at the equilibrium point.
     - (2) Find the quantity of tickets at the equilibrium point.
     - (3) Explain what these values represent with regard to your interest as the supplier and the consumers' interests.

- Write your journal entry for this unit. As usual, before you begin to write, review the material in the unit. Think about how it all fits together. Try to identify what, if anything, is still causing you trouble.

HW3.17 Write your journal entry for Unit 3.

1. Reflect on what you have learned in this unit. Describe in your own words the concepts that you studied and what you learned about them. How do they fit together? What concepts were easy? Hard? What were the main (important) ideas? Give some examples of the main ideas.
2. Reflect on the learning environment for the course. Describe the aspects of this unit and the learning environment that helped you understand the concepts you studied. What activities did you like? Dislike?

3

Name ______________________________ Section __________ Date ______

Unit 4:

LIMITS

Mathematical expressions like "the slope of the curve," "zero sum," "normalized distribution," and "asymptotic" are no longer just mutterings of bearded thinkers who cannot remember to wear socks of the same color. They have become part of the basic vocabulary of business, politics, library management, health care and even social work. One important reason is that mathematical expressions give us a way of thinking about relationships that would otherwise be unavailable to us. Just as your ability to think more complex thoughts was enhanced every time you learned a new word or phrase, so your ability to understand abstract concepts will be enriched when you master mathematical ideas like "limits," "nonlinear," and "exponential growth."

Sheila Tobias, *Succeed with Math*, p. 4,
The College Board, 1987.

OBJECTIVES

1. Examine the limiting behavior of functions.
2. Approximate the values of limits using input–output tables and graphs.
3. Formulate a limit-based definition for continuity.
4. Evaluate limits.

4

OVERVIEW

Limits form the bridge between algebra and calculus. For instance, you can use algebra to find the area of regular shapes, such as rectangles or trapezoids or circles, since you know a formula for each one of these. But what happens if you need to find the area of a shape for which there is no formula, such as the area of an oddly shaped region bounded by two curves? Limits can help you do this and do much more. They are one of the most fundamental building blocks in calculus. You will use them frequently.

Your primary goal in this unit is to develop a conceptual understanding of what a limit is. You will examine the limiting behavior of a function by analyzing input–output tables and by scrutinizing the graph of the function. You will investigate what happens to the output values of a function as the function's inputs get closer and closer to a specified point. You will use your understanding of limit to develop a definition for continuity.

Because forming conjectures concerning the value of a limit can involve doing lots of calculations, you will use ISETL to do many of the computations for you. You will use ISETL to define sequences that approach a specified value from the left and the right, and you will use ISETL to create input–output tables. In addition, you will use your CAS to graph the functions.

As you use the computer remember not to sit passively while the computer does all the work. Think about what's happening. As we mentioned in the Preface, each time you ask the computer to do something, keep in mind the following:

- What you have commanded the computer to do
- Why you asked it to do whatever it is doing
- How it might be doing whatever you have told it to do
- What the results mean

Thinking this way will help you form a mental image associated with the process of a limit. It will help you understand what is happening and help you recognize when to use a limit to solve a problem.

SECTION 1

Limiting Behavior of Functions

Your study of limits begins by thinking about the meaning of the term *approach.* In the first task, you will examine questions such as: What is a sequence of numbers? What does it mean to say that a sequence of numbers gets closer and closer to or "approaches" another number? How can you represent a typical term in a sequence? How can you generate a sequence of numbers that approaches a given value?

Task 4-1: Constructing Sequences of Numbers

1. Examine some equivalent ways to represent a sequence of small numbers.

a. Consider the sequence of numbers in the first column of the chart given below. Complete the unfilled entries in the chart, expressing each of the items in the sequence as a fraction, as a fraction whose denominator is a power of 10, and as a negative power of 10.

Decimal representation	Fraction	Fraction with denominator power of 10	Negative power of 10
0.1	$\frac{1}{10}$		
0.01			
0.001			10^{-3}
0.0001			
0.00001		$\frac{1}{10^5}$	

b. Represent some of the items in the sequence on the real number line.

(1) Mark the location of 0.1 and 0.01.

(2) Describe how you would determine the location of 0.001.

c. Suppose you use the same pattern to continue defining terms in this sequence. Consider the term that has 1 in the 42nd decimal place and zeros everywhere else.

(1) Represent this term as a fraction whose denominator is a power of 10.

4

(2) Represent this term as 10 raised to a negative exponent.

d. It is impossible to write down every member in the extended sequence, because the sequence goes on forever. You can, however, represent a "typical term" in the sequence, since the items in the sequence follow a specific pattern. In particular, the first term has a 1 in the first decimal place, the second term has a 1 in the second decimal place, the third term has a 1 in the third decimal place, and so on. In general, the ith term has a 1 in the ith decimal place and zeros everywhere else, where i is any positive integer.

(1) Represent the ith term in the sequence as a fraction whose denominator is a power of 10.

(2) Represent the ith term in the sequence as 10 raised to a negative exponent.

e. If you continue to define members in this sequence using the same pattern, the items in the sequence get closer and closer to a particular value.

(1) What is this value?

(2) Will the terms in the sequence ever actually reach this value? Explain why or why not.

(3) The sequence is said to be approaching this value *from the right.* What does "from the right" mean in this case?

2. Construct another sequence of numbers.

a. Sequences can be defined using bases other than 10. For instance, the chart given below gives some members of a sequence defined using powers of 2. Complete this chart and then compare this sequence to the one you considered in part 1. Note: A series of three dots indicates that some terms in the sequence are missing. Do not fill in anything in this space.

Decimal representation	Fraction	Fraction with denominator power of 2	Negative power of 2
	$\frac{1}{2}$		
	$\frac{1}{4}$		
	$\frac{1}{8}$		
	$\frac{1}{16}$		
	$\frac{1}{32}$		
⋮	⋮	⋮	⋮
	$\frac{1}{512}$		
⋮	⋮	⋮	⋮
	$\frac{1}{2048}$		

b. Extend this sequence using the same pattern.

(1) What value is the sequence approaching?

(2) Will the sequence ever reach this value?

(3) Is the sequence approaching the value from the left or the right?

4

c. Represent a typical item in this new sequence by an expression—that is, represent the ith term in the sequence by an expression, where i is any positive integer.

(1) Express the ith term in the sequence as a fraction whose denominator is a power of 2.

(2) Express the ith entry in the sequence as 2 raised to a negative exponent.

d. Compare the two sequences you considered above, one of which was defined in terms of powers of 10 and the other in terms of powers of 2.

One of the sequences approaches zero from the right "more rapidly" than the other. Determine which sequence approaches zero more rapidly and explain why. Support your conclusion by locating some of the entries in each sequence on the number line given below. Mark the first three items in the base 10 sequence with a small X and mark the first three items in the base 2 sequence with a small O.

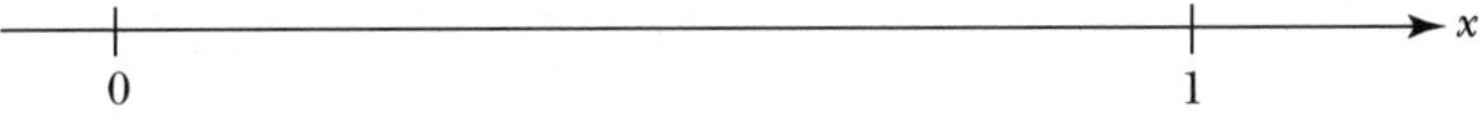

3. The two sequences that you considered above both approach zero from the right. Define two sequences which approach zero *from the left.*

a. Using powers of 10, construct a sequence which approaches zero from the left.

(1) List some of the values in the sequence.

(2) Represent the ith entry in this sequence by an expression, where i is any positive integer.

b. Using powers of 2, construct a sequence which approaches zero from the left.

(1) List some of the terms in the sequence.

(2) Represent the ith entry in the sequence by an expression, where i is any positive integer.

4. Construct sequences that approach a value other than zero.

a. Consider the sequence

$$6.1,\ 6.01,\ 6.001,\ 6.0001, \ldots$$

(1) Assuming the sequence continues on indefinitely, what value does the sequence appear to be approaching? Is the sequence approaching this value from the right or the left?

(2) Represent the ith entry in the sequence by an expression. *Hint:* $n.001$ can be rewritten as $n + 10^{-3}$, where n is an integer.

b. Using powers of 10, construct a sequence that approaches 6 from the left.

(1) List some of the items in this sequence.

(2) Represent the ith entry in this new sequence by an expression, where i is any positive integer.

c. Using powers of 2, construct a sequence that approaches -4 from the left. Represent the ith entry in the sequence, where i is any positive integer.

4

d. Using powers of 2, construct a sequence that approaches −4 from the right. Represent the *i*th entry in the sequence, where *i* is any positive integer.

In the last task, you examined the limiting behavior of a sequence. You explored ways to construct a sequence of numbers that gets closer and closer to a particular value, such as 0, 6 , or −4, without ever actually reaching the value. In general, if a is a fixed real number and the values of x are getting closer and closer to a from the right, you can write

$$x \to a^+$$

where the arrow sign $\to$ is read as "approaches" and the little $^+$ sign indicates that the values of x are greater than—and hence to the right of—a. Similarly, if the values of x are approaching a from the left, you can write

$$x \to a^-$$

Finally, if they are approaching a from both sides—that is, if $x \to a^+$ and $x \to a^-$—you can write

$$x \to a$$

It is important to note that in all three of these cases the arrow notation indicates that the values of x get closer and closer to a, but do not reach a itself.

What you are really interested in, however, is the limiting behavior of a function. In particular, how do the output values of a function behave as the associated inputs approach a particular number? For example, what happens to the value of $f(x) = x + 5$, as the values of x get closer and closer to 6? In other words, does $x + 5$ approach a particular value, as x approaches 6? If so, what? Or, as another example, suppose *DayLight* is a function that returns the length of each day during the year. What happens to the value of *DayLight*(t), as t approaches the vernal equinox? One way to investigate the answers to questions such as these is to graph the given function and observe what's happening. Another way is to create an *input-output table.* You will use both of these approaches in the next task.

Caution! As you examine the limiting behavior of the following functions, keep in mind that there are two "approachings" going on: one by the inputs to the function and one by the corresponding output values of the function. These are different and they happen simultaneously.

Task 4-2: Analyzing the Limiting Behavior of Functions

1. Examine the limiting behavior of $f(x) = x + 5$ as x approaches 6.

a. Use an input–output table to examine the limiting behavior of the function.

(1) Consider the sequence of numbers approaching 6 from the right and the sequence approaching 6 from the left that you defined in Task 4-1, part 4. Calculate the value of $f(x) = x + 5$ at each input and record the result in the *input–output table* given below.

Term # i	Input (x) $6 - 10^{-i}$	Output Value $f(x)$	Input (x) $6 + 10^{-i}$	Output Value $f(x)$
1	5.9	10.9		
2				
3				
4				
5				
	↓ 6^- (from the left)	↓ ☐	↓ 6^+ (from the right)	↓ ☐

(2) The output values appear to approach a particular value as x approaches 6 from the left—that is, as $x \to 6^-$. What value is this? Record your response in the first box in the last row of the input–output table.

This value is called the *left-hand limit* of the function f as x approaches 6. One way to express this mathematically is to write

$$f(x) \to \square \quad \text{as } x \to 6^-$$

or, instead of using the name of the function, you can use the expression that represents the function and write

$$x + 5 \to \square \quad \text{as } x \to 6^-$$

Fill in the boxes with the appropriate value.

4

(3) Next note what happens when you approach 6 from the other side. What value does the function's output values appear to approach as x approaches 6 from the right—that is, as $x \to 6^+$? Record your response in the last row of the input–output table.

This value is called the *right-hand limit* of the function f as x approaches 6. In this case, you write

$f(x) \to \square$ as $x \to 6^+$ or $x + 5 \to \square$ as $x \to 6^+$

where the boxes are filled in with the appropriate value.

(4) You should have concluded that the right- and left-hand limits are the same. If you didn't, retrace your steps and check your calculations. Because the two one-sided limits are equal, this value is called the *limit* of the function f as x approaches 6, and you write:

$f(x) \to \square$ as $x \to 6$ or $x + 5 \to \square$ as $x \to 6$

b. Investigate the limiting behavior of this function again, but this time use a graphic approach.

(1) Consider the graph of $f(x) = x + 5$ for $2 \leq x \leq 10$ given below. Using a pencil and straightedge, indicate the output value corresponding to each of the specified inputs located to the left and right of $a = 6$. (The first association is already marked.)

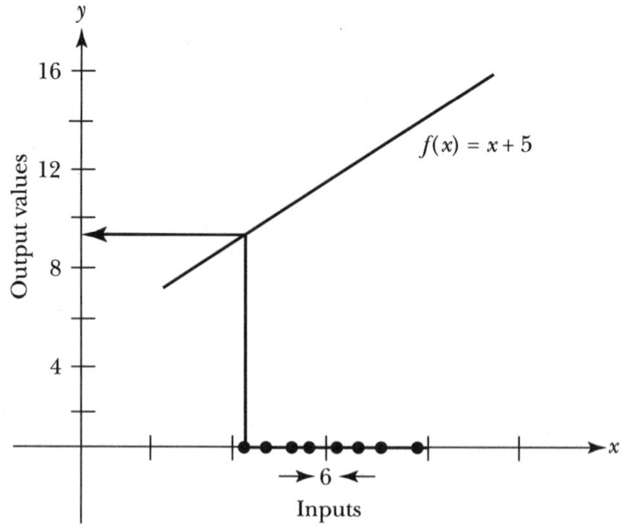

(2) Based on your graph, what can you conclude about the behavior of the output values of f as x approaches 6 from the left and from

the right? Mark the value on the vertical axis. Indicate the limiting behavior of the function as $x \to 6^-$ and as $x \to 6^+$ by placing vertical arrows along the vertical axis in the appropriate directions.

(3) You should have reached the same conclusion using the input–output table and the graph. If you did not, try again.

2. Examine the limiting behavior of a piecewise-defined function near a hole.

Note: Keep in mind that you are interested in the limiting behavior of a function near the specified point, not at the point. Consequently, the function does not have to be defined at the point.

Consider the following function which has a hole at $x = 2$:

$$h(x) = \begin{cases} 2x, & \text{if } x < 2 \\ x^2, & \text{if } x > 2 \end{cases}$$

a. Use an input–output table to analyze the limiting behavior of h as x approaches 2.

Fill in the following table. Define a sequence of numbers that approaches $a = 2$ from the left, called LeftSeq, and a sequence that approaches $a = 2$ from the right, called RightSeq. Evaluate h at each of these inputs. Predict the limiting behavior of h as $x \to 2^-$ and as $x \to 2^+$.

Term # i	LeftSeq (x) 2 − ☐	Output Values $h(x)$	RightSeq (x) 2 + ☐	Output Values $h(x)$
1				
2				
3				
4				
5				
	↓ 2^-	↓ ☐	↓ 2^+	↓ ☐

4

b. Use a graphic approach to analyze the limiting behavior of h as x approaches 2.

On the axes below, sketch the graph of h for $0 \le x \le 4$. Using a pencil and straightedge, indicate on the graph the output values corresponding to several input values immediately to the left of $a = 2$ and several immediately to the right of $a = 2$. On the vertical axis, mark the value h approaches as x approaches 2. Use vertical arrows to indicate the limiting behavior of h as $x \to 2^-$ and $x \to 2^+$.

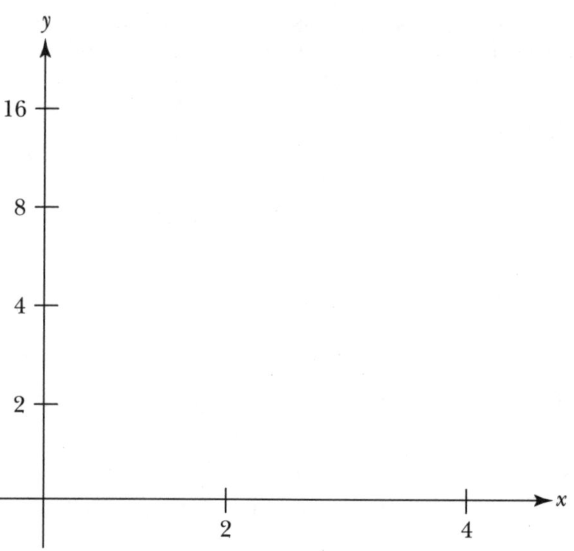

c. Summarize your observations from parts a and b. (The results should be the same.)

$h(x) \to \square$ as $x \to 2^-$ and $h(x) \to \square$ as $x \to 2^+$

Consequently,

$h(x) \to \square$ as $x \to 2$

d. Express the left- and right-hand limits, representing h by the appropriate expression to the left and to the right of $a = 2$.

$\square \to \square$ as $x \to 2^-$

and

$\square \to \square$ as $x \to 2^+$

e. Explain why it is legal to consider the limit of a function at a hole.

3. In this task, you have considered two approaches for examining the limiting behavior of a function: using an input–output table and using the graph. Describe the relationship between these two approaches.

In general, a function f has a limit L as x approaches a, if $f(x)$ can be made to get as close to L as you wish by restricting x to a small interval about a, but excluding a. In this case you write

$$f(x) \to L \quad \text{as } x \to a$$

Two important observations:

- When you refer to the limit of a function as x approaches a, you are referring to how f behaves near a, not at a. In fact, a function does not even have to be defined at $x = a$ for the limit to exist.
- In order for L to be the limit of a function f as x approaches a, the values of $f(x)$ must be getting closer and closer to L as x approaches a from both the left and from the right. The left-hand and right-hand limits must both exist and must both be equal to L.

In the last task, you used two methods to help make a reasonable guess for the value of L:

- You examined an input–output table for f as $x \to a$ from both the left and the right.
- You examined the graph of f for x near a.

In the next task, you are going to use the same methods to estimate the value of the limit of a given function as x approaches a specified value, but this time instead of doing all your work by hand, you will use ISETL to construct an input–output table and your CAS to graph the function near the indicated point.

Task 4-3: Approximating Limits Using the Computer

1. Construct some sequences in ISETL.

   - Work through the following session interactively. Try it, as you read along.
   - Think ISETL! After entering the code, but before pressing the Return key, predict what ISETL will return (if anything).

   a. Construct some *tuples*, or sequences, of integers.

   A tuple in ISETL is a sequence of objects, having a first item, a second item, and so on. It is denoted in ISETL by enclosing the items in the tuple between two square brackets: [and].

   Enter the following tuples in ISETL. Think ISETL!

ISETL Code	Predicted Output	Actual Output
[0..10];		
t := [−4..5];		
t;		

   b. Construct sequences that approach a specified value.

   If m and n are integers, then the successive items in the tuple $[m..n]$ differ by 1. In the case of an input–output table, however, you need to define a tuple of values that approach a given value. Instead of having the jump between successive items remain constant, it must become smaller and smaller. For instance, the items in the tuple LeftSeq get closer and closer to 6 from the left

```
LeftSeq := [5.9, 5.99, 5.999, 5.9999];
```

   while the items in the tuple RightSeq get closer and closer to 6 from the right

```
RightSeq := [6.1, 6.01, 6.001, 6.0001];
```

   The items in LeftSeq can be generated by evaluating the ISETL expression

$$6 - 10**(-i), \quad \text{where } i = 1, 2, 3, 4$$

You can use this expression inside an ISETL *tuple former* to construct LeftSeq as follows:

```
LeftSeq := [ 6 − 10**(−i) : i in [1..4] ];
```

To "Think ISETL!" with this tuple former, think about i looping—or iterating—through the values in the tuple [1..4].

- To get the first item LeftSeq, set $i = 1$ and evaluate the expression 6 − 10**(−1).
- To get the second item in LeftSeq, set $i = 2$ and evaluate the expression 6 − 10**(−2), and so on, until $i = 4$.

Evaluate the following tuple formers. As usual, try to predict the output before pressing the Return key.

ISETL Code	Predicted Output	Actual Output
LeftSeq := [6 − 10**(−i) : i in [1..4]]; LeftSeq; RightSeq := [6 + 10**(−i) : i in [1..4]]; RightSeq;		

a := −4;
 	
b := 2;
n := 5;
 	
LeftSeq := [a − b**(−i) : i in [1..n]];
LeftSeq;
 	
RightSeq := [a + b**(−i) : i in [1..n]];
RightSeq;
	$ *a* is the value the sequence approaches.
	$ *b* is the value of the base.
	$ *n* is the number of terms.	

$ *Note*: To avoid typing in the following block of code, edit the lines in the
$ previous block and re-execute them.
 	
a := 0;
b := 10;
 	
n := 6;
 	
LeftSeq := [a − b**(−i) : i in [1..n]];
LeftSeq;
 	
RightSeq := [a + b**(−i) : i in [1..n]];
RightSeq;
	
	$ Change the value of *a*.
	$ Change the value of the base.
	
	$ Change the number of terms.	

$ *Experiment*. Change the value of *a*. Change the value of the base *b*. Change the $ number of items *n* in the tuple. Observe how these changes affect the values in the tuple.		

4

2. Investigate how you might use the computer to analyze the limiting behavior of a given function.

Analyze $g(x) = (x^3 + 2x^2 + x + 2)/(x + 2)$ as x approaches -2.

a. Create an input–output table in ISETL.

(1) Duplicate the following session, being careful to think about what is happening.

ISETL Code	Predicted Output	Actual Output

$ Define two sequences of numbers approaching $a = -2$ from the left and the right.
 	
a := −2;
 	
b := 10;
 	
n := 5;
 	
LeftSeq := [a − b**(−i) : i in [1..n]];
LeftSeq;
 	
RightSeq := [a + b**(−i) : i in [1..n]];
RightSeq;		

$ Define the function $g(x) = \dfrac{x^3 + 2x^2 + x + 2}{x + 2}$.
 	
g := func (x);
if x /= −2 then
  return (x**3 + 2*x**2 + x + 2)/(x + 2);
end if;
end func;		

$ Create two input–output tables, one corresponding to RightSeq and one to LeftSeq.
 	
for x in LeftSeq do
  writeln x, g(x);
end for;
 	
for x in RightSeq do
  writeln x, g(x);
end for;		

(2) Print a copy of the two input–output tables. Place them in your Activity Guide. Label the columns in the tables LeftSeq (x), Output Values $g(x)$, and so on.

(3) Draw some conclusions about the limiting behavior of $g(x) = (x^3 + 2x^2 + x + 2)/(x + 2)$ as x approaches -2.

b. Graph the function using your CAS.

(1) Graph $g(x) = (x^3 + 2x^2 + x + 2)/(x + 2)$ for x near -2, for instance for $-4 \leq x \leq 0$. Print a copy of the graph and place it in your Activity Guide.

(2) Illustrate graphically that the value of L (which you determined using you input–output tables) is reasonable. As usual, mark L on the vertical axis. Use a pencil and straightedge to indicate the output values corresponding to several input values to the left and to the right of $a = -2$. Indicate the behavior of g as $x \to -2^-$ and as $x \to -2^+$ with vertical arrows.

c. Summarize your observations.

Based on your tables it appears that

$$g(x) \to \boxed{} \text{ as } x \to -2^- \quad \text{and} \quad g(x) \to \boxed{} \text{ as } x \to -2^+$$

Consequently,

$$g(x) \to \boxed{} \text{ as } x \to -2$$

3. Analyze some other functions using your computer.

If your input–output table does not provide enough information about the behavior of a function near $x = a$, then:

- Consider more terms in RightSeq or LeftSeq by increasing the value of n.
- Consider sequences which approach a more rapidly by increasing the value of the base b, say from 2 to 10.
- Increase the precision of the values in the sequence by displaying more digits. (By default, ISETL displays six digits after the decimal point in a real number. To change the number of digits to n, use the precision(n) command. For example, for 10 digits enter precision (10);)

For additional information, refer to the section on creating an input–output table in your ISETL handout.

a. Analyze the limiting behavior of

$$f(x) = \begin{cases} x + 2, & \text{if } x < 0 \\ \sin(x) + 2, & \text{if } x > 0 \end{cases}$$

as $x \to 0$.

(1) Use ISETL to construct input–output tables for f as x approaches 0 from the left and the right. Print copies of the tables and place them in your Activity Guide. Label the columns of the tables and indicate the limiting behavior of f as $x \to 0^-$ and $x \to 0^+$.

(2) Use your CAS to graph f for x near 0. Print a copy of the graph and place it in your Activity Guide. Use a pencil and straightedge to illustrate the limiting behavior of f as $x \to 0^-$ and $x \to 0^+$.

(3) Summarize your observations about the behavior of f for x near 0. Express your result using the arrow ($\to$) notation for limits.

b. Analyze the limiting behavior of

$$f(x) = |x + 1|$$

as $x \to -1$.

(1) Use ISETL to construct input–output tables for f for x near -1. Print copies of the tables and place them in your Activity Guide. Label the columns of the tables and indicate the limiting behavior of f as $x \to -1^-$ and $x \to -1^+$.

(2) Use your CAS to graph f for x near -1. Print a copy of the graph and place it in your Activity Guide. Use a pencil and straightedge to illustrate the limiting behavior of f as $x \to -1^-$ and $x \to -1^+$.

(3) Represent $f(x) = |x + 1|$ by a piecewise-defined function.

$$f(x) = \begin{cases} & \text{if } x < -1 \\ & \text{if } x \geq -1 \end{cases}$$

(4) Summarize your observations about the limiting behavior of f for x near -1. Express your result using the arrow ($\to$) notation for limits and the piecewise representation of f.

Each of the functions you have examined up to this point has had a limit at the indicated point. It is possible, however, that the limit does not exist.

- The value of a limit must be a finite real number. If the output values "explode" as x approach a—that is, if they go off to plus or minus infinity— then the limit does not exist at $x = a$. Consequently, a function does not have a limit at a vertical asymptote.
- The value of the left- and right-hand limits must be the same. If the output values approach one number as x approaches a from the right and a different number as x approaches a from the left, then the limit does not exist at $x = a$. Consequently, a function does not have a limit at a jump.

Consider these two situations in the next task.

Task 4-4: Examining Some Situations Where the Limit Does Not Exist

1. Use input–output tables to show that the following limits do not exist. In each case:

i. Construct the associated input–output tables using ISETL. Place copies of the tables in your Activity Guide.

ii. Explain why the information on your tables supports the claim that the limit does not exist.

a. Show that the limit of

$$p(x) = \frac{x+6}{x-1} \quad \text{as } x \to 1$$

does not exist.

b. Show that the limit of

$$f(x) = \begin{cases} x, & \text{if } x < 0 \\ \cos(x), & \text{if } x \geq 0 \end{cases} \quad \text{as } x \to 0$$

does not exist.

2. Use a graphic approach to show that the following limits do not exist. In each case:

i. Sketch a graph of the function.

ii. Explain why the graph supports the claim that the limit does not exist.

a. Show that the limit of

$$g(x) = \begin{cases} x^2, & \text{if } x \leq 2 \\ 2x + 3, & \text{if } x > 2 \end{cases} \quad \text{as } x \to 2$$

does not exist.

b. Show that the limit of

$$h(x) = \frac{1}{x} \quad \text{as } x \to 0$$

does not exist.

Unit 4 Homework After Section 1

- Complete the tasks in Section One in the Activity Guide. Be prepared to discuss them in class.
- Describe what a limit is in HW4.1.

HW4.1 What is a limit?

Write a one-page essay explaining what a limit is. Include in your explanation a description of how one can find a reasonable estimate for a limit using an input–output table and using a graph. Also include a description of how you can determine when a limit does not exist.

- Consider another notation for limits in HW4.2.

HW4.2 In this section, we used the notation $f(x) \to L$ as $x \to a$ to indicate that the limit of $f(x)$ as x approaches a equals L. Another way to denote this is with the notation

$$\lim_{x \to a} f(x) = L$$

Similarly, writing $f(x) \to LL$ as $x \to a^-$ is equivalent to writing $\lim_{x \to a^-} f(x) = LL$, and $f(x) \to RL$ as $x \to a^+$ is equivalent to $\lim_{x \to a^+} f(x) = RL$, where LL and RL represent the left- and right-hand limits respectively.

1. Give a verbal description of each of the following limits.

a. $\lim_{x \to 0} \dfrac{x}{x^3 - 6x^2 + 5x} = \frac{1}{5}$

b. If $f(x) = \sqrt[3]{2x^2 - 4x + 8}$, then $\lim_{x \to 0} f(x) = 2$.

c. $\lim_{x \to 0^+} \sqrt{x} = 0$

d. Let $g(t) = \begin{cases} 0, & \text{if } t < -10 \\ t + 6.5, & \text{if } t > -10 \end{cases}$

(1) $\lim_{t \to -10^+} g(t) = -3.5$

(2) $\lim_{t \to -10^-} g(t) = 0$

2. Express each of the following limits using the new lim notation.

a. $-2\sin(\theta) \to -2$, as $\theta \to \frac{\pi}{2}$

b. $\dfrac{t^2 - \frac{1}{4}}{t - \frac{1}{2}} \to \frac{3}{4}$, as $t \to \frac{1}{2}$

4

c. $100 \rightarrow 100$, as $x \rightarrow 5$

d. Let $h(r) = \begin{cases} r^2 + 4r + 8, & \text{if } r < -2 \\ -r^2 + 4r + 4, & \text{if } r \geq -2 \end{cases}$

(1) $h(r) \rightarrow 4$, as $r \rightarrow -2^-$
(2) $h(r) \rightarrow -4$, as $r \rightarrow -2^+$

e. $g(x) \rightarrow M$, as $x \rightarrow b^-$

3. Model the responses to each of the following questions as a limit using the lim notation.

a. A patient receives a 200-milligram dose of a drug. Suppose $a(t)$ represents the amount of the drug present in her bloodstream after t hours. What will be the amount of the drug in her bloodstream as the time approaches $2\frac{1}{2}$ hours after the initial dose?

b. A company makes footballs. When n balls are made the average cost per ball is $C(n)$. As the number of footballs produced increases, the average cost per ball drops. What will be the average cost per ball as the number of balls produced nears 1,000,000?

c. Suppose $h(t) = \log(t + 1) + 0.75$, where $t \geq 0$, gives the average height in feet of a particular animal after t years. What will the average height of the animal be as it approaches age 10?

d. Suppose $m(x)$ is the slope of the secant line determined by the points $Q(x, f(x))$ and $P(2,10)$ which lie on the graph of f. What value does the slope of the secant line approach as x gets closer and closer to 2?

- Examine the limiting behavior of some functions in HW4.3.

HW4.3 Use your calculator or computer for the following questions.

1. Use a graphic approach to show that the value for each of the following limits is reasonable. Label the graph carefully.

a. $2x - 3 \rightarrow -7$ as $x \rightarrow -2$

b. $\lim_{x \rightarrow 10} 5 = 5$

c. $x^2 + 1 \rightarrow 5$ as $x \rightarrow 2^-$

d. $\lim_{x \rightarrow \pi} \sin(x) = 0$

2. Use an input–output table to show that the value of each of the following limits is reasonable.

a. $\lim_{x \to 3} 2^x = 8$

b. $4.5 \to 4.5$ as $x \to -3$

c. $\lim_{x \to 0^+} \sqrt{x} = 0$

d. Let $p(x) = \begin{cases} -x + 1, & \text{if } x \leq 0 \\ x^2 + 1, & \text{if } x > 0 \end{cases}$

(1) $\lim_{x \to 0} p(x) = 1$

(2) $\lim_{x \to -2} p(x) = 3$

(3) $p(x) \to 5$ as $x \to 2$

3. Show that each of the following limits does not exist. Use:

i. a graphic approach

ii. an input–output table

a. $\lim_{x \to 0} \dfrac{1}{x^2}$

b. The limit of $h(x)$ as $x \to -5$, where $h(x) = \begin{cases} -3, & \text{if } x \leq -5 \\ 4.5, & \text{if } x > -5 \end{cases}$

• Interpret some graphs and tables in HW4.4.

HW4.4 Find reasonable values for some limits by considering the graph of a function or a table of values.

1. Consider the following graph of function h:

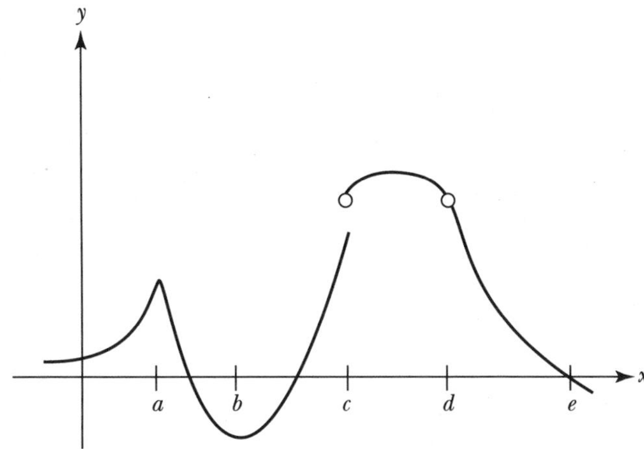

4

Mark the values of L, M, N, P, Q, and R on the vertical axis of the graph of h where:

a. $h(x) \to L$ as $x \to a$

b. $\lim_{x \to b} h(x) = M$

c. $h(x) \to N$ as $x \to c^-$

d. $\lim_{x \to c^+} h(x) = P$

e. $\lim_{x \to d} h(x) = Q$

f. $h(x) \to R$ as $x \to e$

2. Suppose you help individuals prepare their federal income tax returns. Most of your clients manage to file their returns on time by April 15. Those who don't request a 4-month extension. The following graph shows your average workload over the course of a year.

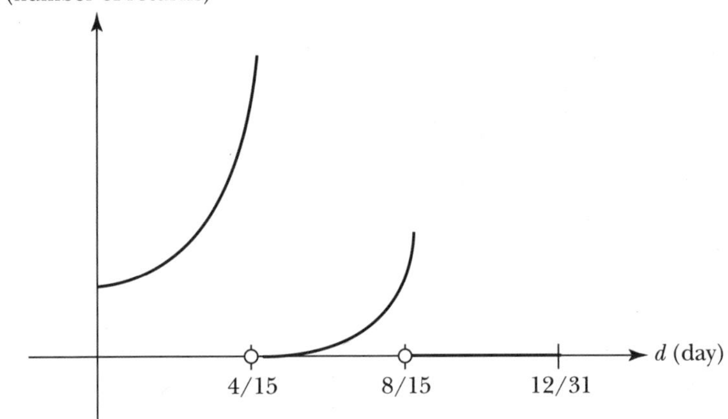

a. Analyze your workload near April 15. Indicate the values of the limits on the vertical axis.

(1) Label the value of $\lim_{d \to 4/15^-} w(d)$.

(2) Label the value of $\lim_{d \to 4/15^+} w(d)$.

(3) Interpret your answers to parts (1) and (2). What do they say about your workload at this time of year?

b. Find and interpret each of the following limits. If the limit does not exist, explain why not.

(1) $\lim_{d \to 12/31^-} w(d)$

(2) $\lim_{d \to 8/15} w(d)$

(3) $\lim_{d \to 6/15} w(d)$

3. Suppose you use ISETL to examine the behavior of a function f for x near -2 and discover that:

x	f(x)	x	f(x)
−2.5	7.8701	−1.5	10.7500
−2.05	8.0600	−1.95	10.5678
−2.005	8.3456	−1.995	10.2345
−2.0005	8.4567	−1.9995	10.0034
−2.00005	8.4999	−1.99995	10.0001

a. From the information that ISETL gives you, what might you conclude about:

(1) the left-hand limit of $f(x)$ as x approaches -2?

(2) $\lim_{x \to 2^+} f(x)$?

(3) the limit of $f(x)$ as x approaches -2?

b. Assume f is defined at $x = -2$. Describe the behavior of the graph of f for x near -2.

- Use limit-based argument to find the slope of a tangent line to a curve. The exercise is a preview of things to come.

HW4.5 The slope of the tangent line to the graph of $f(x) = -x^2 + 2x + 2$ at the point $P(1,3)$ is equal to 0. Use a limit-based argument to show that this is reasonable, and then consider the general case.

1. Consider the following graph of $f(x) = -x^2 + 2x + 2$.

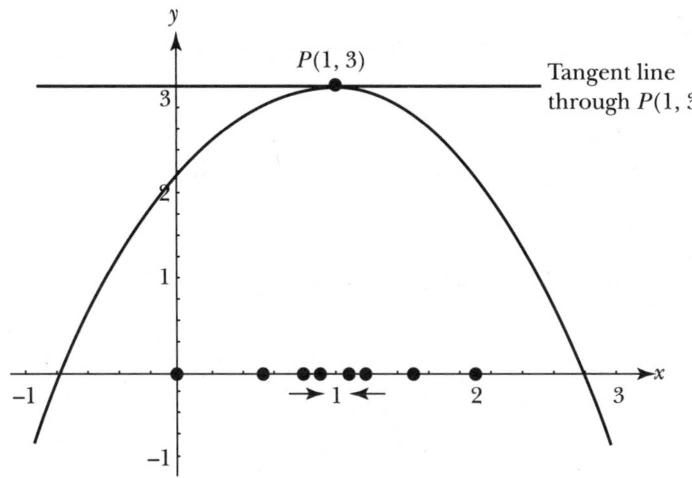

The items in the sequence of points indicated on the horizontal axis approach 1 from the left and from the right. Corresponding to each item in the sequence, there is a point on the graph of f and there is a secant line which passes through the point on the graph and P. Use a graphic approach to examine the limiting behavior of the slopes of these secant lines as $x \to 1$.

a. Consider the sequence of x-values approaching 1 from the left. Use a pencil and straightedge to draw the secant lines determined by P and each point on the graph that corresponds to an item in the sequence. Describe the behavior of the slopes of these secant lines as $x \to 1^-$.

b. Similarly, consider the sequence of x-values approaching 1 from the right. Draw the secant lines determined by P and each point on the graph that corresponds to an item in the sequence. Describe the behavior of the slopes of these secant lines as $x \to 1^+$.

c. Based on your observations, does it appear that the slopes of the secant lines are approaching 0 as $x \to 1$?

2. Use an input–output table to examine the behavior of the slopes of the secant lines as $x \to 1$.

a. Fill in the entries in the following input–output table. Indicate the values the slope of the secant lines appear to be approaching as x approaches 1 from the left and the right.

LeftSeq x	f(x)	Slope of Secant Line Determined by P(1, 3) and Q(x, f(x))	RightSeq x	f(x)	Slope of Secant Line Determined by P(1, 3) and Q(x, f(x))
0	2	$\frac{3-2}{1-0} = 1$	2		
0.5			1.5		
0.75			1.25		
0.9			1.1		
↓ 1^-		↓ ☐	↓ 1^+		↓ ☐

b. Based on the input–output table, does it appear that the slopes of the secant lines are approaching 0 as $x \to 1$?

3. Consider the general case. Examine the limiting behavior of the slopes of the secant lines as $x \to a$, where $-1 < a < 3$.

a. Make a clean sketch of the graph of $f(x) = -x^2 + 2x + 2$ for $-1 \le x \le 3$.

b. Let a be a number between -1 and 3.

(1) Mark a on the horizontal axis and $f(a)$ on the vertical axis.
(2) Mark the point $P(a, f(a))$ on the graph of f.

c. Consider a point close to $x = a$.

(1) Let h be a small number. Mark $a + h$ on the horizontal axis and $f(a + h)$ on the vertical axis.
(2) Mark the point $Q(a + h, f(a + h))$ on the graph of f.

d. Using a straightedge, draw the secant line determined by the points P and Q.

e. Find an expression, in terms of a and h, for the slope of the secant line determined by P and Q.

f. Note that as h gets closer and closer to zero, $a + h$ approaches a. Describe the behavior of the slope of the secant lines determined by P and Q as $h \to 0$.

SECTION 2

Continuity, Limits, and Substitution

The word "continuous" means to go on without interruption. If a function is *continuous* at a point in its domain, you can trace through the graph of the function without lifting your pencil; if it is not, you have to lift your pencil at the place on the graph corresponding to the point.

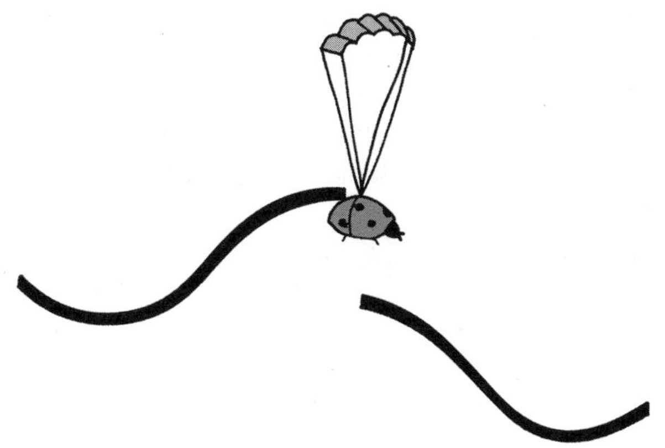

The never-lift-your-pencil description of continuity provides you with a visual way to recognize when a function is continuous at a point. Limits, however, can provide you with a mathematical description of continuity. One way to discover the relationship between continuity and limits is to examine the limiting behavior of a function at a point of *discontinuity* and analyze what goes wrong. This is the approach you will use in the next task. You will then use your observations to develop a limit-based definition of continuity. This new definition will provide you with a straightforward way to evaluate the limit of a continuous function that does not involve constructing input-output tables or sketching the graph.

Before beginning the next task, recall the various types of discontinuities (see HW2.4). A function has one of the following:

- *Removable discontinuity* or *hole* at $x = a$ if the function can be redefined so that it is continuous at $x = a$.
- *Jump discontinuity* at $x = a$ if the graph of the function has a vertical gap at $x = a$.
- *Blowup discontinuity* at $x = a$ if the graph of the function has a vertical asymptote at $x = a$—that is, if the values of the function explode as x approaches a from one side or the other.

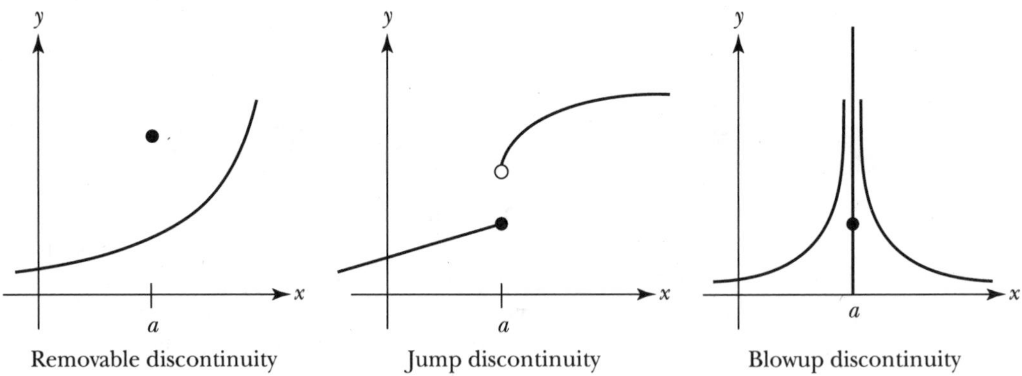

Removable discontinuity Jump discontinuity Blowup discontinuity

Task 4-5: Inspecting Points of Discontinuity

Examine the various types of discontinuities and determine limit-based conditions guaranteeing that these types of discontinuities cannot occur.

Note: When discussing the continuity or discontinuity of a function at a point, the function is assumed to be defined at that point.

1. Examine the limiting behavior of a function near a jump discontinuity.

a. On the axes below, sketch the graph of a function that is defined at $x = a$ and has a jump at $x = a$.

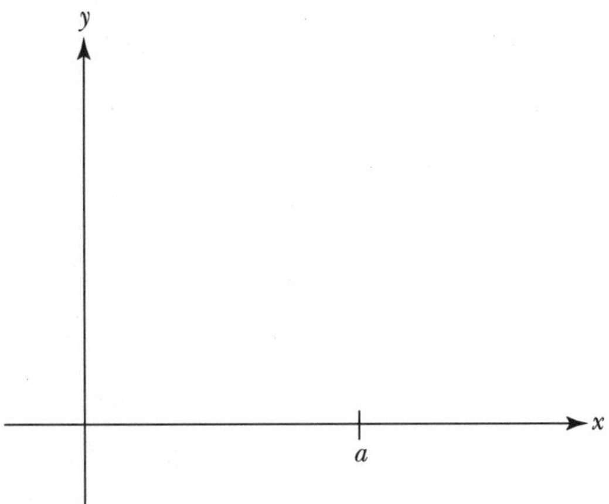

b. On the vertical axis of your graph:

(1) Label $f(a)$.

(2) Label LL, where LL is the left-hand limit of f as x approaches a; that is, $LL = \lim_{x \to a^-} f(x)$.

(3) Label RL, where RL is the right-hand limit of f as x approaches a; that is, $RL = \lim_{x \to a^+} f(x)$.

c. Does the limit of f as x approaches a (from both sides) exist if f has a jump discontinuity at $x = a$? Justify your response.

d. Clearly, a function that has a vertical jump at $x = a$ is not continuous at $x = a$. The goal, however, is to find conditions that guarantee that a function *is* continuous. Based on your observations, what condition on the existence of the limit, and in particular on LL and RL, will eliminate the possibility of a function having a vertical gap at $x = a$?

2. Examine the limiting behavior of a function near a blowup discontinuity.

a. On the axes on the next page, sketch the graph of a function that is defined at $x = a$ and has a vertical asymptote at $x = a$.

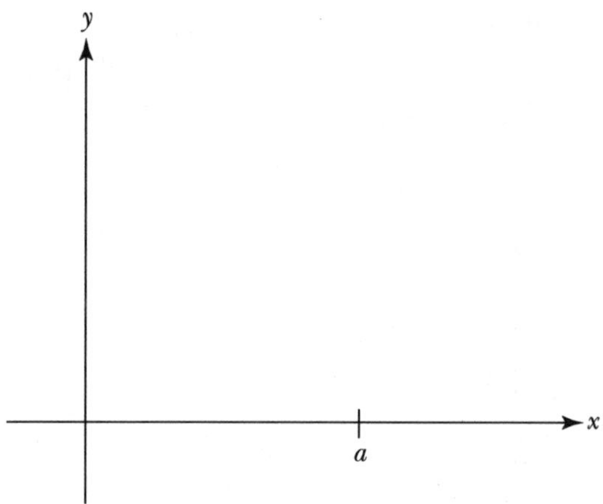

b. Label $f(a)$ on the vertical axis of your graph.

c. Does the limit of f as x approaches a exist if f has a blowup discontinuity at $x = a$? Justify your response.

d. Based on your observations, what condition on $\lim_{x \to a} f(x)$ will guarantee that a function does not have a vertical asymptote at $x = a$?

3. Examine the limiting behavior of a function near a removable discontinuity.

a. On the axes below, sketch the graph of a function that is defined at $x = a$ and has a hole at $x = a$.

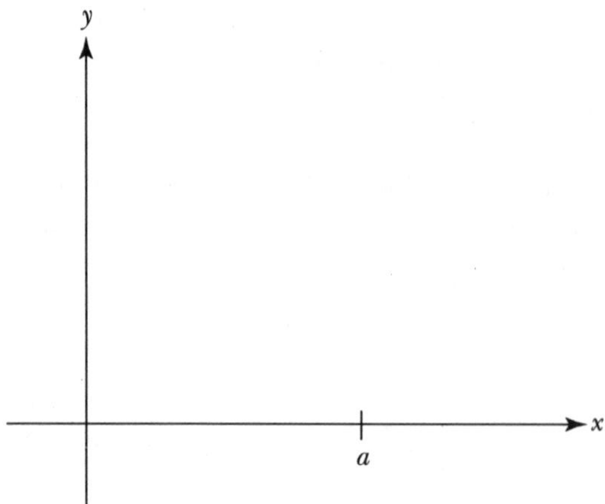

b. On the vertical axis of your graph:

(1) Label $f(a)$.

(2) Label L, where $L = \lim_{x \to a} f(x)$.

c. In the case of a function that is defined at $x = a$ and has a hole at $x = a$, what can you conclude about the relationship between the value of the limit as x approaches a and the value of $f(a)$?

d. Based on your observations, what limit-based condition guarantees that the point $P(a, f(a))$ is connected to the graph? In other words, what must be true about the values of L and $f(a)$ for a function not to have a hole at $x = a$?

What conditions guarantee that a function is continuous at a point? According to your observations in Task 4-5, a function is continuous at a point if and only if:

i. The function is defined at the point. (It doesn't make sense to talk about a function being continuous at a point where it is undefined.)

ii. The limit of the function exists as x approaches the point. (This condition eliminates the possibility of the function having a jump or a blow-up discontinuity.)

iii. The value of the limit equals the value of the function at the point. (This condition eliminates the possibility of the function having a removable discontinuity.)

In other words, f is continuous at $x = a$ if and only if:

i. $f(a)$ exists

ii. $\lim_{x \to a} f(x)$ exists

iii. $\lim_{x \to a} f(x) = f(a)$

The third condition provides you with a quick way to calculate the limit of a continuous function. In particular, it says that if a function is continuous at $x = a$, you can find the value of $\lim_{x \to a} f(x)$ by simply evaluating $f(a)$. This enables you to find the exact value of a limit instead of using an input–output table or a graph to find a reasonable approximation. To use this approach you need to be able to identify if a function is continuous. This is the goal of the next task.

Task 4-6: Identifying Continuous Functions

1. Use the limit-based definition of continuity to examine the continuity of some functions at a point.

For each function ask yourself:

i. Does $f(a)$ exist? If it does, find its value.

ii. Does $\lim_{x \to a} f(x)$ exist? If it does, find a reasonable value for the limit by considering the graph of the function near $x = a$. Sketch a graph of the function by hand or use your CAS. Place the graph in your Activity Guide.

iii. Assuming the answers to i and ii are "yes," does $\lim_{x \to a} f(x) = f(a)$?

If the answer to all three questions is "yes," then the function is continuous at $x = a$. If the answer to any one of them is "no," then f is not continuous at $x = a$. Quit as soon as you encounter the first "no" answer.

a. Determine if $f(x) = \sin(x)$ is continuous at $x = \pi$.

(1) Responses to questions.

(2) Conclusion.

b. $f(x) = \dfrac{1}{\cos(x)}$ at $x = \dfrac{\pi}{2}$

(1) Responses to questions.

(2) Conclusion.

c. $f(x) = \begin{cases} x - 2, & \text{if } x < 4 \\ \sqrt{x}, & \text{if } x \geq 4 \end{cases}$ at $x = 4$

(1) Responses to questions.

(2) Conclusion.

d. $f(x) = \begin{cases} e^x, & \text{if } x \geq 0 \\ x, & \text{if } x < 0 \end{cases}$ at $x = 0$

(1) Responses to questions.

(2) Conclusion.

e. $f(x) = \begin{cases} \dfrac{1}{x}, & \text{if } x \neq 0 \\ 0, & \text{if } x = 0 \end{cases}$ at $x = 0$

(1) Responses to questions.

(2) Conclusion.

f. $f(x) = \begin{cases} x^2, & \text{if } x \neq -2 \\ 5, & \text{if } x = -2 \end{cases}$ at $x = -2$

(1) Responses to questions.

(2) Conclusion.

2. A function is *continuous over an interval I* if it is continuous at each point in the interval. Consider the basic function classes. Identify the intervals where functions in these classes are continuous.

a. The class of constant functions.

Consider an arbitrary constant function $f(x) = c$, where c is a constant. Sketch the general shape of the graph of f. Based on the graph, identify the set of all x-values where a constant function is continuous.

b. The class of linear functions.

Consider an arbitrary linear function $f(x) = mx + b$, where $m \neq 0$. Sketch the general shape of the graph of f. Based on the graph, identify the set of all x-values where a linear function is continuous.

c. The class of parabolic functions.

Consider an arbitrary parabolic function $f(x) = ax^2 + bx + c$, where $a \neq 0$. Sketch the general shape of the graph of f. Based on the graph, identify the set of all x-values where a parabolic function is continuous.

d. The class of power functions.

(1) Let n be an even integer, where $n > 0$. Consider an arbitrary even power function $f(x) = x^n$. Sketch the general shape of the graph of f. Based on the graph, identify the set of all x-values where an even power function is continuous.

(2) Let n be a odd integer, where $n > 0$. Consider an arbitrary odd power function $f(x) = x^n$. Sketch the general shape of the graph of f. Based on the graph, identify the set of all x-values where a odd power function is continuous.

e. The class of logarithmic functions.

Consider an arbitrary exponential function $f(x) = \log_b x$, where $b > 0$ and $b \neq 1$. Sketch the general shape of the graph of f. Based on the graph, identify the set of all x-values where an exponential function is continuous.

f. The general sine function.

Consider an arbitrary sine function $f(x) = a \sin(b(x - c)) + d$, where $a \neq 0$ and $b \neq 0$. Sketch the general shape of the graph of f. Based on the graph, identify the set of all x-values where a sinusoidal function is continuous.

g. The cosine function.

Consider the cosine function $f(x) = \cos(x)$. Sketch the graph of f. Based on the graph, identify the set of all x-values where the cosine function is continuous.

3. In the last part, you identified where functions in the basic function classes are continuous. In addition, any combination of continuous functions is continuous on its domain. In particular, the sum, difference, product, quotient, and composition of continuous functions is continuous on its domain. This enables you to identify where numerous other functions are continuous. For example,

- Every polynomial is continuous for all real numbers, since a polynomial is the sum of power functions, which, in turn, are continuous.
- Every rational function is continuous at all points where the denominator is not equal to zero, since a rational function is the quotient of two polynomials, which, in turn, are continuous.
- Every extended power function ($x^{m/n}$) is continuous on its domain, since it is the composition of two continuous functions.
- A complicated function, such as $f(x) = (3\sin(0.5x) + \log_2(x))/3^x$ is continuous for all $x > 0$, since f is the quotient of two continuous functions. In particular, the numerator of f is continuous for all $x > 0$, since the numerator is a combination of a sinusoidal function (which is continuous for all real numbers) and a logarithmic function (which is continuous for all positive numbers). The denominator is continuous for all real numbers, since every exponential function is continuous for all real numbers. Moreover, the denominator is never 0, since an exponential function is never 0.

Use similar arguments to explain why each of the following functions is continuous on its specified domain.

a. $g(x) = \dfrac{x^2 - 2.4x + 9.5}{\sqrt{x}}$, where $x > 0$.

b. $h(r) = 6\log_{10}(r - 2) - 3r^{10}$, where $r > 2$.

c. $s(t) = \pi + \cos(t)$, where t is any real number.

4

If a function f is continuous at $x = a$, then

$$\lim_{x \to a} f(x) = f(a)$$

This enables you to use *substitution* to find the exact value of the limit of a continuous function by simply evaluating the function at $x = a$. This powerful result makes calculating limits much easier—and certainly more precise—than using input–output tables or graphs.

Because a combination of continuous functions is continuous, you can find the limit of a combination by uncombining the components, finding the limit of each of each component, and then combining the results. In particular, if the limit of f and the limit of g as x approaches a exist, where

$$\lim_{x \to a} f(x) = L \qquad \text{and} \qquad \lim_{x \to a} g(x) = M$$

then the following hold:

- The limit of the sum, difference, or product of two functions is the sum, difference, or product of the limits.

$$\lim_{x \to a} (f(x) \pm g(x)) = \lim_{x \to a} f(x) \pm \lim_{x \to a} g(x) = L \pm M$$
$$\lim_{x \to a} (f(x) \cdot g(x)) = \lim_{x \to a} f(x) \cdot \lim_{x \to a} g(x) = L \cdot M$$

- The limit of a quotient is the quotient of the limits, provided the limit of the denominator is not 0.

$$\lim_{x \to a} \frac{f(x)}{g(x)} = \frac{\lim_{x \to a} f(x)}{\lim_{x \to a} g(x)} = \frac{L}{M}, \quad \text{where } M \neq 0$$

- The limit of the product of a constant and a function is the product of the constant and the function.

$$\lim_{x \to a} c \cdot f(x) = c \cdot \lim_{x \to a} f(x) = c \cdot L$$

Caution! Always interpret the result when you use substitution to calculate a limit. Keep in mind what the value of the limit tells you about the limiting behavior of the function. Remember that the output values of the function approach the value of the limit—which you found using substitution—as the inputs get closer and closer to the specified number; that is, if $\lim_{x \to a} f(x) = L$, then $f(x)$ approaches L as x approaches a from both sides.

Task 4-7: Calculating Limits Using Substitution

Evaluate some limits using substitution and interpret the results. In each exercise:

a. Determine if the function is continuous at $x = a$. Justify your conclusion.

b. If it is continuous, find the limit using substitution.

c. If you can find the limit using substitution, interpret the result. Explain what the value of the limit tells you about the limiting behavior of the function.

1. Consider $\lim_{x \to 1.5} (6x^2 - 2x + 5)$.

a. Is $f(x) = 6x^2 - 2x + 5$ continuous at $x = 1.5$? Justify your conclusion.

b. If the answer to part a is "yes," use substitution to evaluate the limit.

c. If you evaluated the limit using substitution, interpret the result.

2. Consider $\lim_{t \to 0} \log_2 t$.

a. Is $f(t) = \log_2 t$ continuous at $t = 0$? Justify your conclusion.

b. If the answer to part a is "yes," use substitution to evaluate the limit.

c. If you evaluated the limit using substitution, interpret the result.

3. Consider $\lim_{x \to 106} 16.78$.

a. Is $f(x) = 16.78$ continuous at $x = 106$? Justify your conclusion.

b. If the answer to part a is "yes," use substitution to evaluate the limit.

c. If you evaluated the limit using substitution, interpret the result.

4. Consider $\lim_{r \to -2} \frac{r^3 - 1}{r + 6.5}$.

a. Is $h(r) = (r^3 - 1)/(r + 6.5)$ continuous at $r = -2$? Justify your conclusion.

b. If the answer to part a is "yes," use substitution to evaluate the limit.

c. If you evaluated the limit using substitution, interpret the result.

5. Consider $\lim_{x \to -4} \sqrt{x}$.

a. Is $f(x) = \sqrt{x}$ continuous at $x = -4$? Justify your conclusion.

b. If the answer to part a is "yes," use substitution to evaluate the limit.

c. If you evaluated the limit using substitution, interpret the result.

6. Consider $\lim_{n \to -1} (n^{10} - 8n^8 + 6n^6 - 4n^4 + 2n^2 - 100)$.

a. Is $f(n) = n^{10} - 8n^8 + 6n^6 - 4n^4 + 2n^2 - 100$ continuous at $n = -1$? Justify your conclusion.

b. If the answer to part a is "yes," use substitution to evaluate the limit.

c. If you evaluated the limit using substitution, interpret the result.

7. Consider $\lim_{t \to \pi} 6\left(\sin\left(\frac{t}{2}\right) + \cos(t)\right)$.

a. Is $f(t) = 6(\sin(t/2) + \cos(t))$ continuous at $t = \pi$? Justify your conclusion.

b. If the answer to part a is "yes," use substitution to evaluate the limit.

c. If you evaluated the limit using substitution, interpret the result.

8. Consider $\lim_{x \to 1} (e^{(x-1)} \ln(x))$.

a. Is $g(x) = e^{(x-1)} \ln(x)$ continuous at $x = 1$? Justify your conclusion.

b. If the answer to part a is "yes," use substitution to evaluate the limit.

c. If you evaluated the limit using substitution, interpret the result.

9. Consider $\lim_{x \to 4} \frac{1/2^x}{\log_4 x}$.

4

a. Is $q(x) = (1/2^x/\log_4 x)$ continuous at $x = 4$? Justify your conclusion.

b. If the answer to part a is "yes," use substitution to evaluate the limit.

c. If you evaluated the limit using substitution, interpret the result.

Unit 4 Homework After Section 2

- Complete the tasks in Section Two in the Activity Guide. Be prepared to discuss them in class.
- Analyze some piecewise-defined functions in HW4.6.

HW4.6

1. Consider the piecewise-defined function f represented by the following graph:

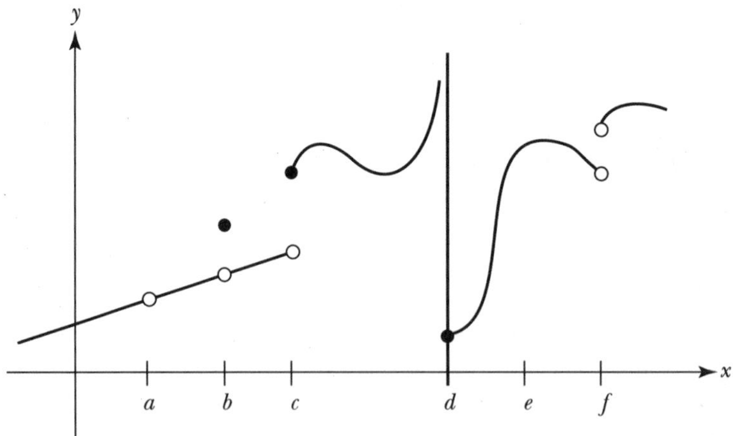

a. Identify the points on the horizontal axis where f is not continuous. For each point, determine which part of the limit-based definition for continuity f fails to satisfy: (i) f is not defined at the point, or (ii) f is defined, but the limit does not exist at the point, or (iii) f is defined and the limit exists, but the value of the limit is not equal to the value of the function at the point.

b. Classify each of the function's points of discontinuity as a jump, blowup, or removable discontinuity.

c. List all the intervals where f is continuous.

d. Mark each of the following statements true or false. Justify each response.

(1) $\lim_{x \to a^-} f(x) = \lim_{x \to a^+} f(x)$

(2) $\lim_{x \to b} f(x) = f(b)$

(3) $\lim_{x \to c^-} f(x) = \lim_{x \to c^+} f(x)$

(4) $\lim_{x \to d} f(x) = f(d)$

(5) $\lim_{x \to e} f(x) = f(e)$

(6) $\lim_{x \to f^-} f(x) < \lim_{x \to f^+} f(x)$

2. Analyze the following piecewise-defined functions.

a. Let $g(x) = \begin{cases} -x, & \text{if } x \leq 0 \\ x^2, & \text{if } 0 < x < 4 \\ 2x, & \text{if } x \geq 4 \end{cases}$

(1) Sketch a graph of the function.
(2) Identify any points in the domain where the function is discontinuous. Classify each point of discontinuity.
(3) List all the intervals where the function is continuous.
(4) Evaluate each of the following limits, if possible. If the limit does not exist, explain why.

(a) $\lim_{x \to -6} g(x)$

(b) $\lim_{x \to 0} g(x)$

(c) $\lim_{x \to 4} g(x)$

(d) $\lim_{x \to 6} g(x)$

(e) $\lim_{x \to 3.5} g(x)$

(f) $\lim_{x \to 4^+} g(x)$

b. Let $h(x) = \begin{cases} \sin(x), & \text{if } x \leq 0 \\ \ln(x), & \text{if } x > 0 \end{cases}$

(1) Sketch a graph of the function.
(2) Identify any points in the domain where the function is discontinuous. Classify each point of discontinuity.
(3) List all the intervals where the function is continuous.
(4) Evaluate each of the following limits, if possible. If the limit does not exist, explain why.

(a) $\lim_{x \to -\pi} h(x)$

(b) $\lim_{x \to 0} h(x)$

(c) $\lim_{x \to 1} h(x)$

(d) $\lim_{x \to 0^-} h(x)$

c. $h(x) = \begin{cases} 2, & \text{if } x < 4 \\ 3, & \text{if } x \geq 4 \end{cases}$

(1) Sketch a graph of the function.
(2) Identify any points in the domain where the function is discontinuous. Classify each point of discontinuity.
(3) List all the intervals where the function is continuous.
(4) Evaluate each of the following limits, if possible. If the limit does not exist, explain why.

(a) $\lim_{x \to -7} h(x)$ **(c)** $\lim_{x \to 7} h(x)$

(b) $\lim_{x \to 4^+} h(x)$ **(d)** $\lim_{x \to 4} h(x)$

- Think about three important theorems concerning continuous functions and limits in HW4.7, HW4.8, and HW4.9. Explain why their statements are reasonable.

HW4.7 According to the *Intermediate Value Theorem for Continuous Functions:*

If f is continuous over the closed interval $[a,b]$, then f takes on every value between $f(a)$ and $f(b)$. In other words, if y is any value between $f(a)$ and $f(b)$, then there exists an x-value c between a and b such that $f(c) = y$.

1. Apply the Intermediate Value Theorem for Continuous Functions to a specific example. Consider $f(x) = 2x + 1$, where $-2 \leq x \leq 4$.

a. Sketch the graph of f for $-2 \leq x \leq 4$. Label -2 and 4 on the horizontal axis and $f(-2)$ and $f(4)$ on the vertical axis.

b. In order to apply the theorem, f must be continuous over the given closed interval. Explain why f is continuous on the closed interval $[-2,4]$.

c. The theorem claims that if y is any value between $f(-2)$ and $f(4)$, there exists an input value c between -2 and 4 such that $f(c) = y$. Apply the theorem for $y = 6$; that is, find a value for c between -2 and 4 such that $f(c) = 6$. Label c and $f(c)$ on the graph you sketched in part a and indicate the relationship between c and $f(c)$.

d. Apply the theorem for $y = -2$; that is, find a value for c between -2 and 4 such that $f(c) = -2$. Label c and $f(c)$ on the graph you sketched in part a and indicate the relationship between c and $f(c)$.

2. Explain why the Intermediate Value Theorem for Continuous Functions makes sense. Support your explanation with an appropriate diagram; that is, on a pair of axes:

a. Label a and b on the horizontal axis, where $a < b$.

b. Sketch the graph of a squiggly function f that is continuous over the closed interval $[a,b]$.

c. Label $f(a)$ and $f(b)$ on the vertical axis.

d. Pick a value between $f(a)$ and $f(b)$ and label it y.

e. Show that the conclusion of the Intermediate Value Theorem for Continuous Functions holds; that is, show that there exists an x-value c between a and b such that $f(c) = y$. Label c on the horizontal axis.

3. For a given value of y between $f(a)$ and $f(b)$, is it possible for there to be more than one choice of c between a and b such that $f(c) = y$? If so, draw a diagram supporting your conclusion. If not, explain why not.

4. Apply the Intermediate Value Theorem to some real-life situations. For each of the following situations:

i. Model the scenario with a graph.

ii. Illustrate the conclusion on your graph.

a. *Scenario:* The light turns green, and you step on the gas pedal in your car. Fifteen seconds later, you level off your speed at 60 MPH.

Conclusion: According to the Intermediate Value Theorem, there exists a time in the 15-second time interval when you are going 28 MPH.

b. *Scenario:* You walk back and forth in front of a motion detector for 20 seconds. You vary your velocity and maintain a distance of 0.5 to 8.5 meters from the detector.

Conclusion: According to the Intermediate Value Theorem, there exists a time in the 20-second time interval when you are 5.25 meters from the detector.

c. *Scenario:* You have the flu. Over a three-day period, your temperature fluctuates between 99.8° and 103.2°.

Conclusion: According to the Intermediate Value Theorem, there exists a time in the three-day period when your temperature is 100°.

5. Use the Intermediate Value Theorem to show that the equation $x^3 - 3x^4 - 2x^3 - x + 1 = 0$ has a solution between 0 and 1.

4

HW4.8 According to the *Max-Min Theorem for Continuous Functions*:

If f is continuous over the closed interval $[a,b]$, then f takes on both an absolute minimum and an absolute maximum on $[a,b]$. In other words:

- There exists a value m between a and b such that $f(m) \leq f(x)$ for all x between a and b; that is, f has an absolute minimum at $x = m$.
- There exists a value M between a and b such that $f(x) \leq f(M)$ for all x between a and b; that is, f has an absolute maximum at $x = M$.

1. Apply the Max-Min Theorem for Continuous Functions to a specific example. Consider $f(x) = x^2 - 1$, where $-2 \leq x \leq 3$.

   a. Sketch the graph of f for $-2 \leq x \leq 3$. Label -2 and 3 on the horizontal axis and $f(-2)$ and $f(3)$ on the vertical axis.

   b. In order to apply the theorem, f must be continuous over the given closed interval. Explain why f is continuous on the closed interval $[-2,3]$.

   c. The theorem claims that f has an absolute minimum on $[-2,3]$—that is, there exists a value m between -2 and 3 such that $f(m) \leq f(x)$ for all x between -2 and 3. Find m. Label m and $f(m)$ on the graph you sketched in part a.

   d. The theorem claims that f has an absolute maximum on $[-2,3]$—that is, there exists a value M between -2 and 3 such that $f(x) \leq f(M)$ for all x between -2 and 3. Find M. Label M and $f(M)$ on the graph you sketched in part a.

2. Explain why the Max-Min Theorem for Continuous Functions makes sense. Support your explanation with an appropriate diagram. On a pair of axes:

   a. Label a and b on the horizontal axis, where $a < b$.

   b. Sketch the graph of a (squiggly) function f which is continuous over the closed interval $[a,b]$.

   c. Show that there exists an input value m between a and b such that $f(m) \leq f(x)$ for all x between a and b. Label m and $f(m)$ on your diagram.

   d. Show that there exists an input value M between a and b such that $f(x) \leq f(M)$ for all x between a and b. Label M and $f(M)$ on your diagram.

3. The Max-Min Theorem for Continuous Functions states that every function f which is continuous over a closed interval $[a,b]$ takes on both an absolute minimum and an absolute maximum on $[a,b]$. If you want to determine where the absolute extrema exist, what values would you examine? In other words, which values in the interval $[a,b]$ would be candidates for the absolute maximum and absolute minimum?

4. Is it possible for there to be more than one choice of m between a and b such that $f(m) \leq f(x)$ for all x between a and b? If so, draw a diagram supporting your conclusion. If not, explain why not. Similarly, is it possible for there to be more than one choice of M between a and b such that $f(x) \leq f(M)$ for all x between a and b? If so, draw a diagram supporting your conclusion. If not, explain why not.

5. Apply the Max-Min Theorem to some real-life situations. For each of the following situations:

i. Model the scenario with a graph.

ii. Label the absolute extrema for your model.

a. You work an 8-hour shift at a pretzel factory. At the start of your shift, your production rate is low, but it continues to increase as you settle into a routine. Two hours before the end of the shift, you start thinking about what you are going to do after work, and your production rate decreases until it's time to quit.

b. You're home alone watching a scary movie, on a dreary night. Each time a scary part comes on, your heart rate increases dramatically and then returns to normal when the scary part is over. The movie is 117 minutes long, and there are 7 scenes that frighten you.

c. You create a distance-versus-time graph using a motion detector. Starting 7.75 meters from the detector, you walk toward the detector for 6.5 seconds. You walk faster and faster for the first 4 seconds, and then slower and slower for the next 2.5 seconds. You stop at the half-meter mark.

HW4.9 According to the *Sandwich Theorem:*

If f, g, and h are three functions, where for all x near some real number c

$$f(x) \leq g(x) \leq h(x)$$

and

$$\lim_{x \to c} f(x) = \lim_{x \to c} h(x) = L$$

then

$$\lim_{x \to c} g(x) = L$$

In other words, whenever the graph of g is "sandwiched" between the graphs of f and h for all x near some real number c, and f and h have the same limit as x approaches c, then g also has the same limit.

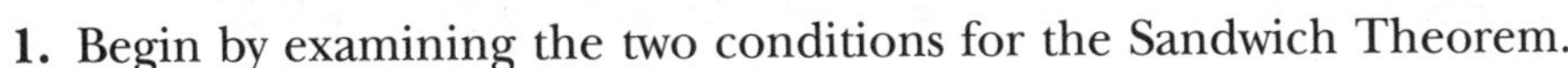

1. Begin by examining the two conditions for the Sandwich Theorem.

a. The first condition states that f, g, and h are three functions, where $f(x) \leq g(x) \leq h(x)$ for all x near c.

(1) Illustrate what this means graphically as follows: Choose an arbitrary value for c on the horizontal axis. For x near c, sketch graphs of functions f, g, and h which satisfy the necessary condition. Label the graphs f, g, and h.

(2) Describe the relationship among the graphs of f, g, and h for x near c.

b. The second condition states that $\lim_{x \to c} f(x) = \lim_{x \to c} h(x) = L$. Consequently, the graphs of f and h "merge" as x gets close to c. Explain why this is the case. Support your explanation with an appropriate diagram.

2. Apply the Sandwich Theorem to a specific example. In particular, use the Sandwich Theorem to find $\lim_{x \to 0} x \sin(1/x)$.

a. Explain why you cannot evaluate $\lim_{x \to 0} x \sin(1/x)$ directly.

b. In order to use the Sandwich Theorem you need to show that you can "sandwich" the graph of $g(x) = x \sin(1/x)$ for x near 0 between the graphs of two functions f and h, where f and h have the same limit as x approaches 0. Consider $f(x) = -|x|$ and $h(x) = |x|$.

(1) Show that the hypotheses of the theorem hold.

(a) Show that $f(x) \leq g(x) \leq h(x)$ for all x near 0 by sketching, on one pair of axes, graphs of f, g, and h for x near 0. *Note:* Use your CAS to sketch the graph of g.

(b) Use a graphic approach to show that $\lim_{x \to 0} f(x) = \lim_{x \to 0} h(x) = 0$.

(2) Find $\lim_{x \to 0} x \sin(1/x)$.

3. Explain why the Sandwich Theorem makes sense in general. Support your explanation with an appropriate diagram. On a single pair of axes:

a. Label c on the horizontal axis and L on the vertical axis.

b. For x near c, sketch graphs of three arbitrary functions f, g, and h where

- $f(x) \leq g(x) \leq h(x)$ for all x near c
- $\lim_{x \to c} f(x) = \lim_{x \to c} h(x) = L$

c. Use the graph to show that the conclusion of the Sandwich Theorem makes sense; that is, use a graphic approach to show that in this case it is reasonable to assume that $\lim_{x \to c} g(x)$ equals L.

SECTION 3

More Limits

At this point, you should have a solid conceptual understanding of what a limit is. You know how to analyze the limiting behavior of a function using an input–output table and a graphic approach. You can recognize when the limit does not exist, and you can use substitution to evaluate the limit of a continuous function. But what if a function is not continuous? What if it has a hole, jump or vertical asymptote? In this section you will investigate how to calculate the limit at a hole, how to evaluate one-sided limits at a jump, and how to use limits to locate horizontal asymptotes. In addition, you will explore how to use *infinite limits* to determine the behavior of a graph near a vertical asymptote.

Holes

Based on your observations in Section One you know that the limit exists at a point where a function has a hole. Direct substitution does not work, however, because either the function is undefined at $x = a$—in which case $f(a)$ does not exist—or the function is defined at $x = a$, but $f(a) \neq L$. One way around this dilemma is to remove the discontinuity and to construct a new function that has the same values as f, except at $x = a$ where its value is L. Since a limit only cares about the values of a function near a, not at $x = a$, the limiting behavior of the new function and the limiting behavior of f will be the same as x approaches a. Moreover, since the new function is continuous at $x = a$, you can use substitution to find its limit.

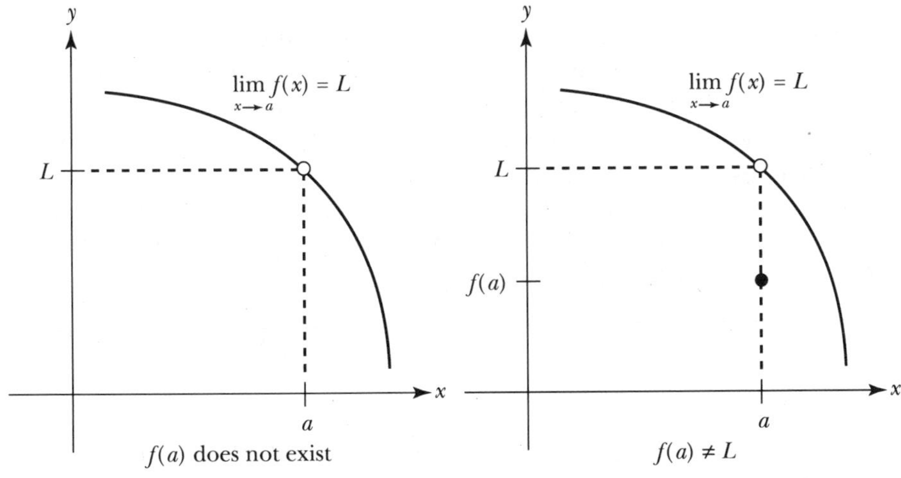

Consider, for example, the rational function

$$f(x) = \frac{3x^2 + 9x + 6}{5x^2 + 4x - 1} = \frac{3(x+1)(x+2)}{(x+1)(5x-1)}, \quad \text{where } x \neq -1 \text{ and } x \neq \frac{1}{5}$$

Because the $(x + 1)$ terms cancel, f has a hole at $x = -1$. The $\lim_{x \to -1} f(x)$ exists, but you cannot calculate the limit by substitution since f is not defined at $x = -1$. However, you can construct a new function,

$g(x) = 3(x+2)/(5x-1)$, which has the same values as f, except it is continuous at $x = -1$. Consequently, the two functions have the same limit as $x \to -1$, and you can use substitution to evaluate the limit of g.

$$\lim_{x \to -1} \frac{3x^2 + 9x + 6}{5x^2 + 4x - 1} = \lim_{x \to -1} \frac{3(x+1)(x+2)}{(x+1)(5x-1)} \qquad \text{Factor}$$

$$= \lim_{x \to -1} \frac{3(x+2)}{(5x-1)} \qquad \text{Cancel } (x+1) \text{ term}$$

$$= \frac{3(-1+2)}{-5-1} \qquad \text{Evaluate by substitution}$$

$$\lim_{x \to -1} \frac{3x^2 + 9x + 6}{5x^2 + 4x - 1} = -\frac{1}{2} \qquad \text{Simplify}$$

The value of the limit gives the y-coordinate of the hole. Since $f(x)$ is approaching $-\frac{1}{2}$ as x approaches -1, f has a hole at $P(-1, -\frac{1}{2})$.

Vertical Asymptotes

Limits can be used to analyze the behavior of a function at a vertical asymptote. Consider the rational function given above. f has a vertical asymptote at $x = \frac{1}{5}$ since the $(5x-1)$ term in the denominator does not cancel with a like term in the numerator. Consequently, the values of f explode as x approaches $\frac{1}{5}$. To determine how the function behaves near the asymptote, you can consider the sign of $f(x)$ on either side of $x = \frac{1}{5}$. For instance, when x approaches $\frac{1}{5}$ from the left, x is less than $\frac{1}{5}$ and the value of $f(x)$ is negative since all the factors in $f(x)$ are positive except $(5x-1)$.

$$f(x) = \frac{3(x+1)(x+2)}{(x+1)(5x-1)} = \frac{+ \cdot + \cdot +}{+ \cdot -} = \frac{+}{-} < 0$$

Although f does not have a finite limit at $x = \frac{1}{5}$, you can write

$$\lim_{x \to \frac{1}{5}^-} f(x) = -\infty$$

to indicate that f is exploding in a negative sense as x approaches $\frac{1}{5}$ from the left. In this case, the "limit" is called an *infinite limit.* You can use a similar approach to show that as x approaches $\frac{1}{5}$ from the right, $f(x)$ is always positive since all the factors in $f(x)$ are positive. In this case, f is exploding in a positive sense and you can indicate this by writing

$$\lim_{x \to \frac{1}{5}^+} f(x) = +\infty$$

End-Points of Subdomains

Limits can also be used to analyze the behavior of a piecewise-defined function at the common end-point of two adjacent subdomains. In this case, you evaluate the left- and right-hand limits separately and compare the values; that is, you calculate the left-hand limit by substituting into the expression that defines the function to the left of the end-point. Similarly, you calculate the right-hand limit by substituting into the expression that defines the function to the right of the end-point. There are several possible outcomes: (1) If the left- and right-hand limits are equal, but the function is not defined, then it has a hole. (2) If the left- and right-hand limits are equal and the function is defined, then it is continuous. (3) If the left- and right-hand limits are not equal, the function has a jump. (4) If either the left- and/or the right-hand limit does not exist, the function has a vertical asymptote. These four possibilities are based on ideas you considered in earlier sections. They should seem reasonable to you.

Before beginning the next task, read through this introduction again and then use the ideas as you do the task.

Task 4-8: Using Limits to Investigate Functions

1. First consider some general situations.

a. Suppose f has a vertical asymptote at $x = a$. For each of the following cases, sketch the graph of f for x near a.

(1) $\lim_{x \to a^-} f(x) = +\infty$ and $\lim_{x \to a^+} f(x) = +\infty$

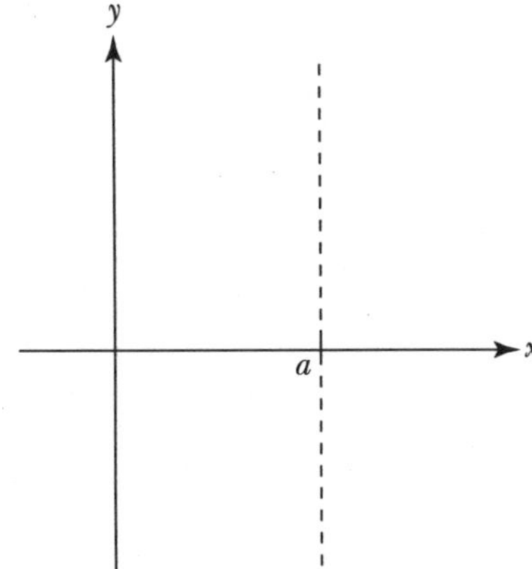

4

(2) $\lim_{x \to a^-} f(x) = -\infty$ and $\lim_{x \to a^+} f(x) = -\infty$

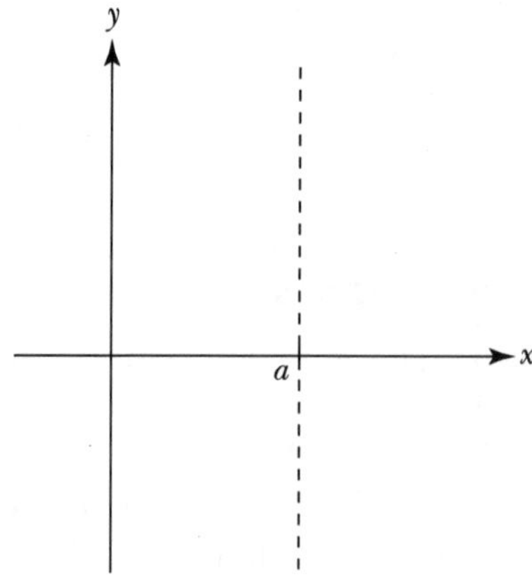

(3) $\lim_{x \to a^-} f(x) = +\infty$ and $\lim_{x \to a^+} f(x) = -\infty$

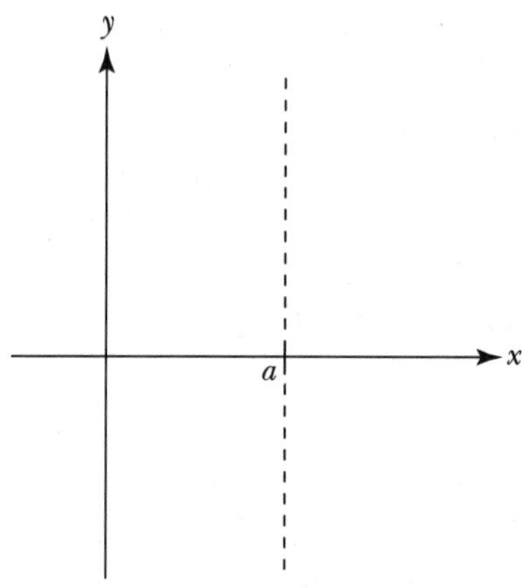

(4) $\lim_{x \to a^-} f(x) = -\infty$ and $\lim_{x \to a^+} f(x) = +\infty$

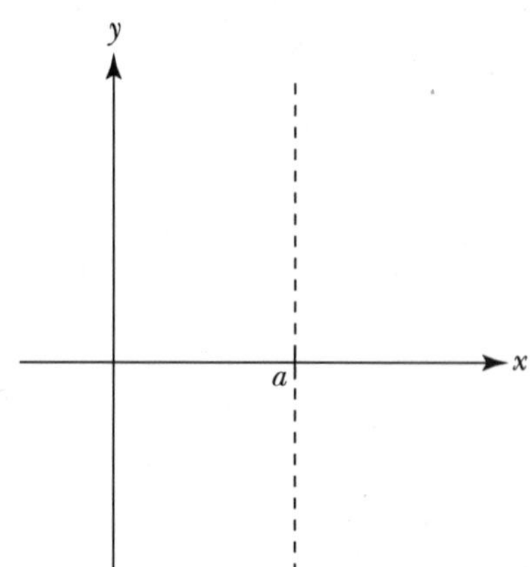

b. Suppose f is a piecewise-defined function where $x = a$ is the common end-point of two adjacent subdomains. For each of the following lists of conditions, sketch the graph of a function satisfying the conditions for x near a.

(1) $\lim_{x \to a^-} f(x) = \lim_{x \to a^+} f(x)$, but $f(a)$ does not exist.

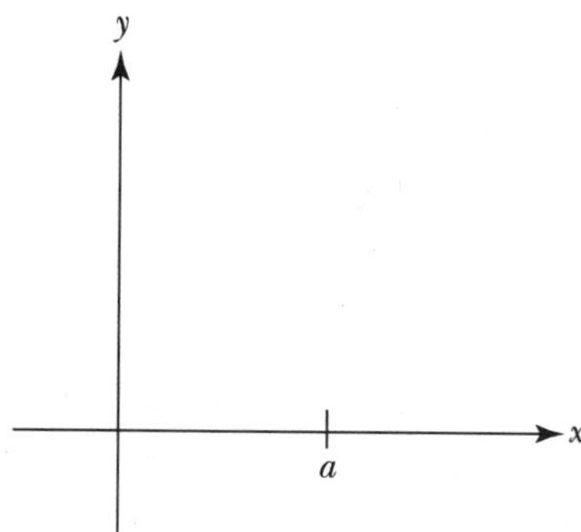

(2) $\lim_{x \to a^-} f(x) = \lim_{x \to a^+} f(x)$ and $f(a)$ exists.

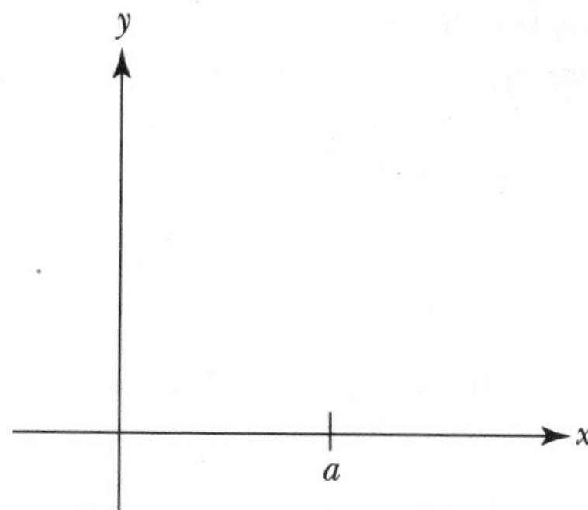

(3) $\lim_{x \to a^-} f(x) \neq \lim_{x \to a^+} f(x)$ and $f(a)$ exists.

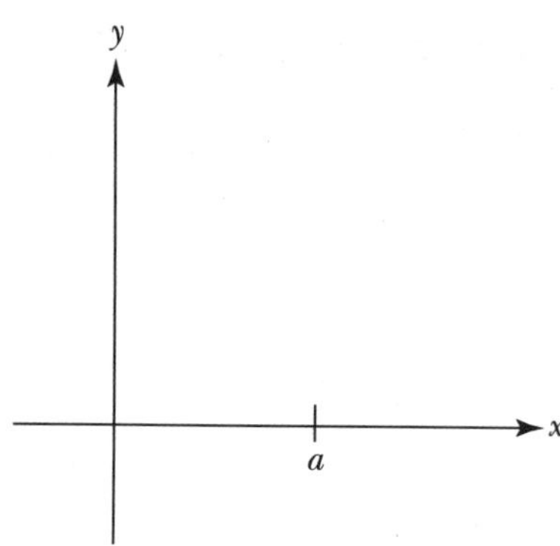

(4) $\lim_{x \to a^-} f(x)$ does not exist, $\lim_{x \to a^+} f(x)$ does exist, and f is defined at $x = a$.

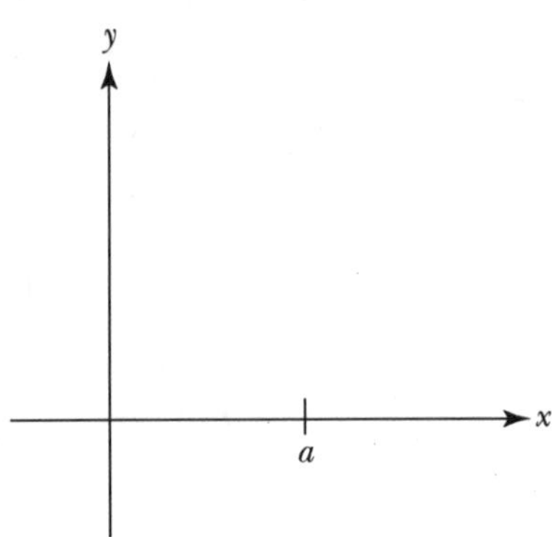

2. Use limits to analyze the behavior of some rational functions (without sketching the graph).

For each value of a, determine if the given function f has a hole, vertical asymptote, or is continuous at $x = a$.

- If f has a hole at $x = a$, evaluate the limit of $f(x)$ as $x \to a$. Find the (x,y)-coordinates of the hole.
- If f has a vertical asymptote, describe the behavior of the function near the asymptote using one-sided limits.
- If f is continuous, evaluate the limit of $f(x)$ as $x \to a$.

a. Consider $f(x) = \dfrac{x^2 + x - 6}{x - 2}$.

(1) Analyze the behavior of f for x near 2.

(2) Analyze the behavior of f for x near -2.

b. Consider $f(x) = \dfrac{x^2 + 6x + 5}{x^2 + 4x - 5}$.

(1) Analyze the behavior of f for x near 1.

(2) Analyze the behavior of f for x near -5.

(3) Analyze the behavior of f for x near 0.

c. Consider $f(x) = \dfrac{-x^3 + x}{x^5 - x}$.

(1) Analyze the behavior of f for x near 2.

(2) Analyze the behavior of f for x near -1.

(3) Analyze the behavior of f for x near 0.

(4) Analyze the behavior of f for x near 1.

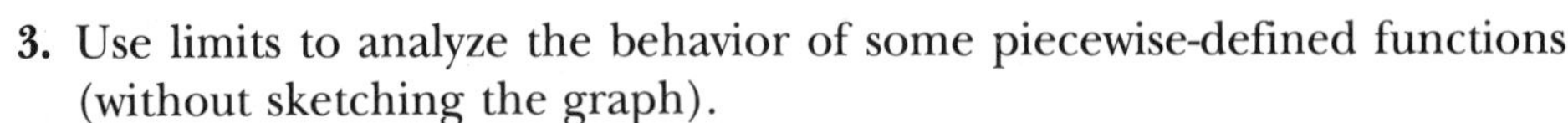

3. Use limits to analyze the behavior of some piecewise-defined functions (without sketching the graph).

For each value of a, find the left- and right-hand limits as $x \to a$. Interpret the results by determining if the given function has a hole, a jump, or is continuous at $x = a$.

a. Consider $f(x) = \begin{cases} \sin(x), & \text{if } x < 0 \\ 2^x, & \text{if } 0 \le x < 2 \\ x^2, & \text{if } x > 2 \end{cases}$

(1) Analyze the behavior of f for x near 0.

(a) Find $\lim_{x \to 0^-} f(x)$.

(b) Find $\lim_{x \to 0^+} f(x)$.

(c) Interpret your results.

(2) Analyze the behavior of f for x near 2.

(a) Find $\lim_{x \to 2^-} f(x)$.

(b) Find $\lim_{x \to 2^-} f(x)$.

(c) Interpret your results.

b. Consider $f(x) = \begin{cases} \sqrt{x^2 - 1}, & \text{if } x < -1 \\ \frac{1}{3}x + \frac{1}{3}, & \text{if } -1 \le x < 2 \\ (x-3)^2, & \text{if } x > 2 \end{cases}$

(1) Analyze the behavior of f for x near -1.

(a) Find $\lim_{x \to -1^-} f(x)$.

(b) Find $\lim_{x \to -1^+} f(x)$.

(c) Interpret your results.

(2) Analyze the behavior of f for x near 2.

(a) Find $\lim_{x \to 2^-} f(x)$.

(b) Find $\lim_{x \to 2^+} f(x)$.

(c) Interpret your results.

4

Thus far you have restricted your attention to thinking about how a function's output values behave as its inputs approach a finite number, such as 3, or -100, or π. Another possibility is to examine how $f(x)$ behaves as the x-values explode; that is, what happens as $x \to -\infty$ or as $x \to +\infty$? If $f(x)$ approaches a finite value, then the function has a *horizontal asymptote.*

For example, consider once again the rational function

$$f(x) = \frac{3x^2 + 9x + 6}{5x^2 + 4x - 1} = \frac{3(x+1)(x+2)}{(x+1)(5x-1)}, \quad \text{where } x \neq -1 \text{ and } x \neq \tfrac{1}{5}$$

You know that f has a hole at $P(-1, -\frac{1}{2})$ and a vertical asymptote at $x = \frac{1}{5}$. Moreover, based on the graph of f, it appears that the graph levels off as x get larger and larger in both the positive and negative directions, and hence that the output values are getting closer and closer to a particular number as the value of x explodes. In other words, it appears that f has a horizontal asymptote. The question is: where?

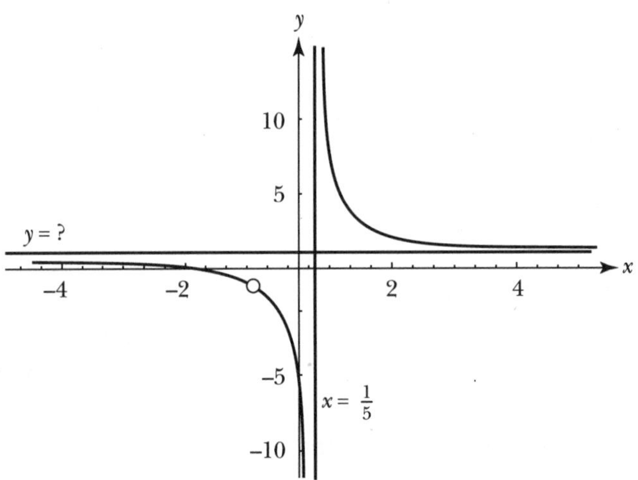

You could use an input–output table to find a reasonable approximation for this horizontal asymptote. However, if you could evaluate

$$\lim_{x \to -\infty} \frac{3x^2 + 9x + 6}{5x^2 + 4x - 1} \quad \text{and} \quad \lim_{x \to +\infty} \frac{3x^2 + 9x + 6}{5x^2 + 4x - 1}$$

you would know the exact value. The next task will guide you through the process of how to do this.

Task 4-9: Using Limits to Locate Horizontal Asymptotes

Obviously, you cannot evaluate a limit such as

$$\lim_{x \to -\infty} \frac{3x^2 + 9x + 6}{5x^2 + 4x - 1}$$

by substituting $-\infty$ for x. One way around this is to find an equivalent expression for the rational function, where each term in the new expression is either a constant or has the form c/x^n, where n is a positive integer and c is a constant. For instance, multiplying the numerator and the denominator of the function given above by $1/x^2$, gives you an equivalent expression where each term has the desired form:

$$\frac{3x^2 + 9x + 6}{5x^2 + 4x - 1} = \frac{(3x^2 + 9x + 6) \cdot \frac{1}{x^2}}{(5x^2 + 4x - 1) \cdot \frac{1}{x^2}} = \frac{3 + (9/x) + \frac{6}{x^2}}{5 + (4/x) - \frac{1}{x^2}}.$$

You can then take the limit of this equivalent expression as x approaches $\pm\infty$. To do so, you need to understand how the constant function c and the function c/x^n behave as x approaches $\pm\infty$.

1. Suppose c is a constant. Examine the limiting behavior of the constant function c as x approaches $+\infty$. In particular, use a graphic approach to show that

$$\lim_{x \to -\infty} c = c \qquad \text{and} \qquad \lim_{x \to +\infty} c = c$$

a. Sketch the graph of a constant function where c is an arbitrary constant.

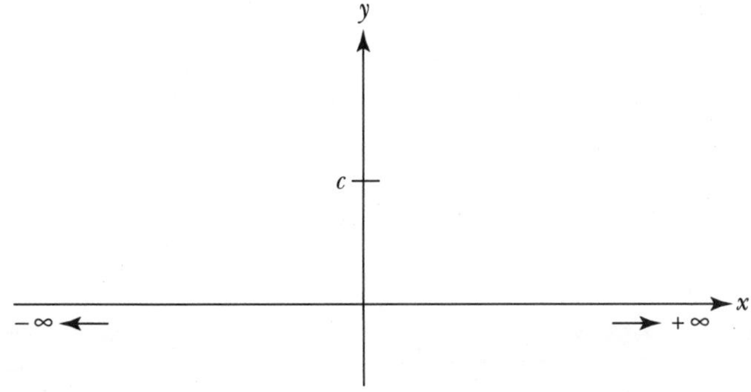

b. Indicate some output values of the constant function as x gets larger and larger in the negative direction, and as x gets larger and larger in the positive direction. Show that the output value equals c, no matter how big or small x gets.

4

2. Suppose n is a positive integer and c is a constant. Examine the limiting behavior of c/x^n as x approaches $+\infty$.

a. Consider $n = 1$. Use a graphic approach to show that it is reasonable to conclude that

$$\lim_{x \to -\infty} \frac{1}{x} = 0 \quad \text{and} \quad \lim_{x \to \infty} \frac{1}{x} = 0$$

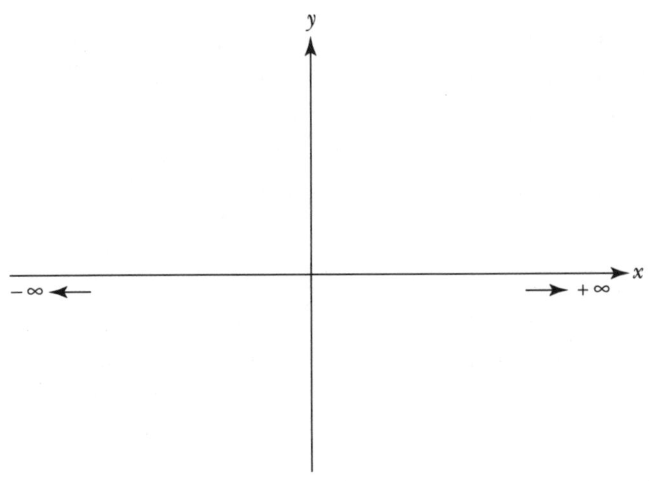

b. Consider $n = 3$. Use input–output tables to show that it is reasonable to conclude that

$$\lim_{x \to -\infty} \frac{1}{x^3} = 0 \quad \text{and} \quad \lim_{x \to \infty} \frac{1}{x^3} = 0$$

x	$\frac{1}{x^3}$
−1	
−10	
−100	
−1,000	
−10,000	

x	$\frac{1}{x^3}$
1	
10	
100	
1,000	
10,000	

c. Generalize your observations. Explain why it is reasonable to assume that

$$\lim_{x \to \infty} \frac{c}{x^n} = 0 \qquad \text{and} \qquad \lim_{x \to -\infty} \frac{c}{x^n} = 0$$

where n is a positive integer and c is a constant.

3. Work through an example. Show that

$$\lim_{x \to \pm\infty} \frac{3x^2 + 9x + 6}{5x^2 + 4x - 1} = \frac{3}{5}$$

and consequently that $f(x) = (3x^2 + 9x + 6)/(5x^2 + 4x - 1)$ has a horizontal asymptote at $y = \frac{3}{5}$.

a. First, find an equivalent expression for

$$\frac{3x^2 + 9x + 6}{5x^2 + 4x - 1}$$

where each term is a constant or has the form c/x^n. One way to do this is to divide each term in the numerator and in the denominator by the highest power of x that appears in the given function, which in this example is x^2. Try it.

4

b. Next, take the limit of the equivalent expression as $x \to \pm\infty$, noting that $c \to c$ and $c/x^n \to 0$ as $x \to \pm\infty$, where n is a positive integer and c is a constant.

4. Evaluate the limits of some rational functions as $x \to \pm\infty$ and locate the horizontal asymptotes. Use the approach developed in the previous example.

a. $\displaystyle \lim_{x \to \infty} \frac{2x^2 - 5}{3x^2 - x + 7}$

b. $\displaystyle \lim_{x \to -\infty} \frac{-0.5 + 1.25x - x^3}{0.75 + 5x^4}$

c. $\displaystyle \lim_{x \to \infty} \frac{\frac{1}{2}x + \frac{2}{3}x^3}{x - \frac{1}{2}x^2 + \frac{1}{3}x^3}$

d. $\lim_{x \to \infty} \left(7 + \frac{3}{x}\right)$

e. $\lim_{x \to \pm\infty} \dfrac{\frac{3}{x^{10}} - \frac{7}{x^6} + 90}{\frac{4}{x^7} - 10}$

5. The limit of a rational function as $x \to +\infty$ may not exist and, hence, a rational function does not necessarily have a horizontal asymptote.

a. Consider a specific example

$$f(x) = \frac{x^3 - 6x^2 + 3}{2x^2 - 9x - 13}$$

Show that f does not have a horizontal asymptote by showing that $\lim_{x \to \pm\infty} f(x)$ does not exist.

4

b. Give two other examples of rational functions which do not have horizontal asymptotes.

c. Describe how can you recognize when a rational function does not have a horizontal asymptote by looking at the expression representing the function.

As you probably suspected, limits can be evaluated using your CAS. Before beginning the next task, determine how to use your CAS to evaluate the limit of a function represented by an expression and by a piecewise-defined function. Determine how to evaluate one- and two-sided limits as x approaches a finite value and as x approaches $\pm\infty$.

Task 4-10: Evaluating Limits Using Your CAS

1. For each of the following functions:

i. Use your CAS to evaluate the specified limits. Write the results in your Activity Guide.

ii. Interpret the results. What does the value of the limit tell you about the behavior of the function near $x = a$?

iii. Use your CAS to graph the function. Check that the values of the limits returned by your CAS are reasonable and that your interpretations of the results makes sense.

a. Let $f(x) = \sqrt[3]{x^3 - 3x^2 + 6.5}$.

(1) Find $\lim_{x \to 1} f(x)$. Interpret the result.

(2) Find $\lim_{x \to 4} f(x)$. Interpret the result.

b. Let $g(x) = \sin(x)/x$.

(1) Find $\lim_{x \to 0^-} \sin(x)/x$ and $\lim_{x \to 0^+} \sin(x)/x$. Interpret the results.

(2) Find $\lim_{x \to -\infty} \sin(x)/x$ and $\lim_{x \to +\infty} \sin(x)x$. Interpret the results.

c. Let $h(t) = \begin{cases} 4e^{-2t}, & \text{if } t < 0 \\ 4\cos(t), & \text{if } 0 \le t < \pi \\ \sin(2t), & \text{if } t \ge \pi. \end{cases}$

(1) Find $\lim_{t \to -\infty} h(t)$. Interpret the result.

(2) Find $\lim_{t \to 0^-} h(t)$ and $\lim_{t \to 0^+} h(t)$. Interpret the results.

(3) Find $\lim_{t \to \pi^-} h(t)$ and $\lim_{t \to \pi^+} h(t)$. Interpret the results.

2. Consider

$$g(r) = \frac{-7.5 - 5r}{-10.14 - 6.76r + 1.5r^2 + r^3}$$

Analyze g using your CAS. Record your results.

a. Use limits to find any horizontal asymptotes of g.

4

b. Find any vertical asymptotes of g. Use limits to determine the shape of the graph of g near each vertical asymptote.

c. Find the x-coordinate of any holes in the graph of g. Use limits to find the y-coordinate of each hole.

Unit 4 Homework After Section 3

- Complete the tasks in Section Three in the Activity Guide. Be prepared to discuss them in class.
- Evaluate some limits in HW4.10.

HW4.10 Use pencil and paper to find each of the following limits, if it exists. If it does not exist, explain why.

1. Let $f(x) = \dfrac{2x^2 + 6x + 4}{x^2 - x - 2}$.

a. $\lim_{x \to -1} f(x)$ **c.** $\lim_{x \to 0} f(x)$

b. $\lim_{x \to 2} f(x)$ **d.** $\lim_{x \to \infty} f(x)$

2. Let $f(x) = \dfrac{10x^6 + 3x^3 - x^2 + 7}{5x^6 + 8x^5 - 1}$.

a. $\lim_{x \to 1} f(x)$ **c.** $\lim_{x \to -\infty} f(x)$

b. $\lim_{x \to 0} f(x)$

3. Let $g(x) = \dfrac{2}{x^2 - 1}$.

a. $\lim_{x \to 0} g(x)$ **c.** $\lim_{x \to -1^-} g(x)$

b. $\lim_{x \to 1^+} g(x)$ **d.** $\lim_{x \to \infty} g(x)$

4. Let $f(x) = \begin{cases} x^3, & \text{if } x < 2 \\ x + 6, & \text{if } 2 \le x < 4 \\ \dfrac{x^3}{3x^3 + 4}, & \text{if } x \ge 4. \end{cases}$

a. $\lim_{x \to 2} f(x)$ **c.** $\lim_{x \to 4^-} f(x)$

b. $\lim_{x \to 3} f(x)$ **d.** $\lim_{x \to 4^+} f(x)$

5. Let $g(x) = \begin{cases} 2^x, & \text{if } x < 0 \\ \dfrac{2x^3 - 4x^2 + x - 1}{-8x^3 - 6x - 1}, & \text{if } x \geq 0. \end{cases}$

a. $\lim_{x \to -2} g(x)$ **c.** $\lim_{x \to 1} g(x)$

b. $\lim_{x \to 0} g(x)$ **d.** $\lim_{x \to \infty} g(x)$

6. Let $h(x) = \begin{cases} \dfrac{x^2 + x - 12}{x + 4}, & \text{if } x < 2 \\ \dfrac{9x^4 - 3x^3 + 6}{-3x^4 - 8x^2 + 7}, & \text{if } x \geq 2. \end{cases}$

a. $\lim_{x \to -4} h(x)$ **c.** $\lim_{x \to \infty} h(x)$

b. $\lim_{x \to 1} h(x)$

7. Let $f(x) = \begin{cases} \dfrac{x^2 - 1}{x - 1}, & \text{if } x < 2 \\ \dfrac{x^4 - x^3 + 7}{2x^4 - 4x^3 + 5}, & \text{if } x \geq 2. \end{cases}$

a. $\lim_{x \to 1} f(x)$ **c.** $\lim_{x \to 2^+} f(x)$

b. $\lim_{x \to 2^-} f(x)$ **d.** $\lim_{x \to -\infty} f(x))$

- Use limits to analyze some situations in HW4.11.

HW4.11 Consider the following situations.

1. Suppose $l(t) = 5 \log_2(t + 1)$ gives the length in feet of a certain poisonous snake after t years. Snakes of this type typically live 8 years. Find the length a typical snake is approaching toward the end of its lifetime.

2. Tests have indicated that the amount of a drug in milligrams remaining in a person's bloodstream after h hours is given by

$$a(h) = \frac{23 - h}{4 + h^2}$$

4

Find and interpret the following limits:

b. $\lim_{h \to 3.5^-} a(h)$ **b.** $\lim_{h \to \infty} a(h)$

3. Joe is always trying new software packages. He has kept track of how long it takes him to feel comfortable with each one. Based on these data, he knows that average number of hours it will take him to learn the nth new package is given by $c(n) = 4 + 6/n$.

a. What is the longest amount of time it has taken him to become familiar with a new package?

b. Joe acquires more and more software packages. As the years pass, what is the average time it takes him to learn a package?

4. Suppose the population of Hometown, USA in thousands t years from now is given by

$$P(t) = \frac{25t^2 + 125t + 200}{2t^2 + 5t + 40}$$

a. What is the current population?

b. What will the population be 13 years from now?

c. What will the population be in the long run?

d. Describe how the population of Hometown changes as time goes by.

5. Suppose you write a best-selling book and your royalties in millions of dollars after x years total

$$P(x) = \frac{4x^2}{x^2 + 4}$$

a. How much will you earn after $2\frac{1}{2}$ years?

b. How much will you earn in the long run?

- Write your journal entry for this unit. As usual, before you begin to write, review the material in the unit. Think about how it all fits together. Try to identify what, if anything, is still causing you trouble.

HW4.12 Write your journal entry for Unit 4.

1. Reflect on what you have learned in this unit. Describe in your own words the concepts that you studied and what you learned about them. How do they fit together? What concepts were easy? Hard? What were the main (important) ideas? Give some examples of the main ideas.

2. Reflect on the learning environment for the course. Describe the aspects of this unit and the learning environment that helped you understand the concepts you studied. What activities did you like? Dislike?

Name ______________________________ Section _________ Date _____

Unit 5:

DERIVATIVES AND INTEGRALS: FIRST PASS

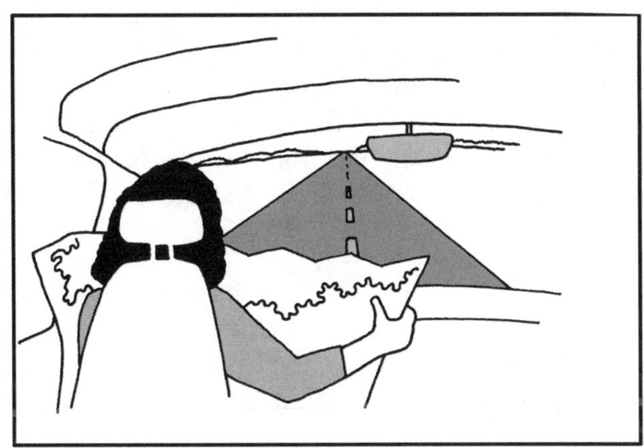

Every natural phenomenon, from quantum vibrations of subatomic particles to the universe itself, is a manifestation of change. Developing organisms change as they grow. Populations of living creatures, from viruses to whales, vary from day to day or from year to year. Our history, politics, economics, and climate are subject to constant, and often baffling, changes. Some changes are simple: the cycles of the season, the ebb and flow of the tides. Others seem more complicated: economic recessions, outbreaks of diseases, the weather. All kinds of changes influence our lives. ... The traditional approach to the mathematics of change can be summed up in one word: calculus.

Ian Stewart, "Change." In Lynn Steen (Ed.): *On the Shoulders of Giants: New Approaches to Numeracy*, National Academy Press, 1990, 183–184.

OBJECTIVES

1. Develop a definition for the derivative.
2. Analyze the behavior of the derivative. Find its domain. Represent it by a graph and an expression.
3. Examine the relationship between a function and its derivatives.
4. Develop a definition for the definite integral.
5. Examine the relationship between a function and its associated accumulation function.
6. Consider the connection between the derivative and the definite integral.

OVERVIEW

This unit provides a first pass at the two basic concepts underlying the study of calculus—the derivative and the definite integral—and the connection between these two ideas.

Actually, you already know a lot about derivatives, under the guise of rates of change and slopes of tangent lines. The problem is that you do not know how to find the actual value of a derivative. For example, you cannot use the two-point formula for slope (which you learned in algebra) to find the slope of the tangent line. Why? Because in the case of a tangent line, you only know one point, namely the point where the tangent line touches the graph. You can use limits, however, to get around this problem.

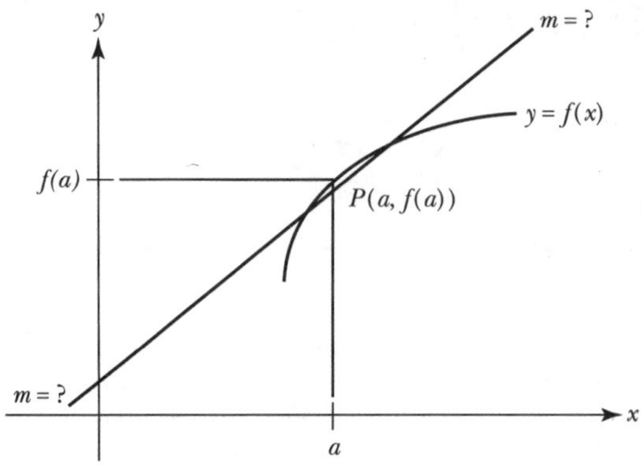

Figure 1.

The idea is to construct a sequence of secant lines that get closer and closer the tangent line. You can calculate the slopes of these lines, since they intersect the graph at two points. You can then find the value of the slope of the tangent line by taking the limit of the slopes of the secant lines as the secant lines approach the tangent line.

The second major concept you will study in this unit is the definite integral. Whereas derivatives are associated with rates of change, integrals correspond to the accumulation of quantities. Integrals can be used, for example, to find the area of an oddly shaped region, such as the shaded region in Figure 2. You know formulas for finding the area of regular shapes, such as rectangles, triangles, and circles, but there isn't a formula for find-

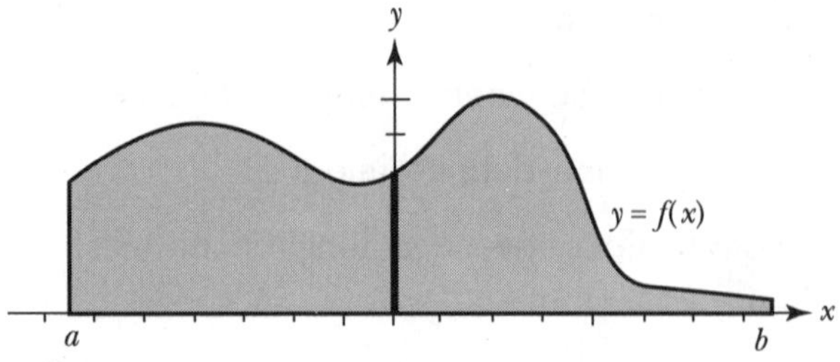

Figure 2.

ing the area of the region that lies under the graph of $y = f(x)$ and over the closed interval $[a,b]$. You can, however, approximate this area by covering it with shapes, whose areas you can calculate, such as rectangles. You can then find the exact value for the area by making the approximations become more and more precise as you take the limit. This is the same approach you will use to find the slope of the tangent line: First approximate the value you want by applying a formula you already know, and then find the exact value by making the approximations become more exact as you take the limit. You will use this approach again and again throughout your study of calculus.

The objectives of this unit are to develop a conceptual understanding of the fundamental concepts, derivative and definite integral, and to make a first pass at examining how these two important ideas are related. Understanding what these ideas mean will help you recognize where they apply. Exploring ways to evaluate them directly and how to use them in a variety of situations is the focus of the next volume.

SECTION 1

The Derivative

In Unit 1 you developed a conceptual understanding of the notion of the tangent line to a curve at a point P. You observed that if you repeatedly zoom in and magnify the portion of the curve containing P, the curve "straightens out" and merges with the tangent line to the curve at P. You noted that the tangent line does not exist at a point where the function has a sharp peak or dip—that is, it does not exist at a *cusp*—because no matter how close you zoom in to a point of this type, the cusp is always there and the graph of the curve never straightens. In HW1.13 and HW4.5, you approximated the value of the slope of the tangent line at a given point by finding the slope of the secant line determined by the given point and a point nearby.

What does all this have to do with derivatives? The value of the derivative of a function at a given point represents the value of the slope of the tangent line to the curve at that point. It also gives the rate of change of the function at the point. In other words,

$$\begin{aligned}&\text{derivative of } f \text{ at } x = a\\ &\qquad \equiv \text{slope of the tangent line to the graph of } f \text{ at } P(a,f(a))\\ &\qquad \equiv \text{rate of change of } f \text{ at } x = a\end{aligned}$$

The question, as you know, is how can you find the exact value? In the next task, you will use limits to find the slope of the tangent line for a specific function. In the following task, you will develop a definition for derivative by formalizing the approach that you develop in Task 5-1. Note: Before beginning the next task, review the ideas developed in homework problems HW1.13 and HW4.5.

Task 5-1: Examining an Example

Consider $f(x) = x^2 + 3$ at $x = 1$. Use a graphic and a numeric approach to examine the limiting behavior of the slopes of the secant lines determined by $P(1,4)$. Based on your observations, find a reasonable value for the slope of the tangent line at $x = 1$. Use a symbolic approach to find the actual value of the slope. Show that the three approaches—graphic, numeric, and symbolic—yield the same result.

1. Use a graphic approach to examine the limiting behavior of the slopes of the secant lines through $P(1,4)$.

   a. Consider the graph of f given below. The sequence of points indicated on the horizontal axis approach 1 from the left and from the right. Mark the points on the graph that correspond to the items in the sequence. Use a pencil and straightedge to draw the secant lines determined by $P(1,4)$ and each of these points.

   Note: Secant lines are not line segments. They go on indefinitely. Extend each line through the two points which determine it.

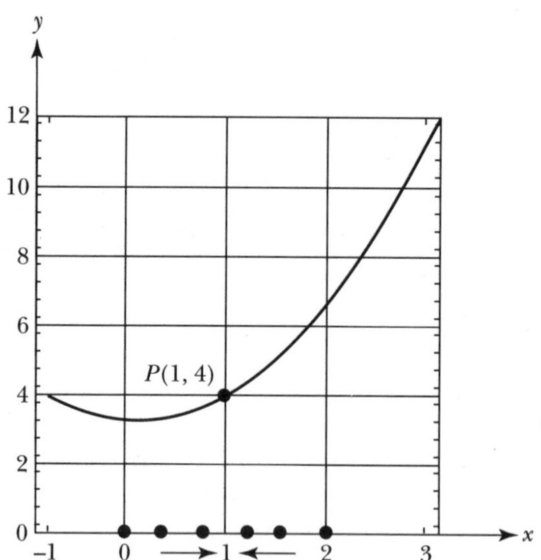

   b. Based on the behavior of the secant lines, find a reasonable value for the slope of the tangent line to the graph of $f(x) = x^2 + 3$ at $x = 1$. *Caution:* The two axes have different scales.

2. Try this again, only this time use a numeric approach to examine the limiting behavior of the slopes of the secant lines through $P(1,4)$.

a. Run the following ISETL session. Think carefully about what is going on here.

ISETL Code	Predicted Output	Actual Output

\$ Define the function $f(x) = x^2 + 3$.
 	
f := func (x);
    return x**2 + 3.
end func;
 	
\$ Set $a = 1$.
 	
a := 1;		

\$ Define two sequences of values, one approaching 1 from the left and one from the right. n := 5; LeftSeq := [a – 10**(–i) : i in [1..n]]; LeftSeq; RightSeq := [a + 10**(–i) : i in [1..n]]; RightSeq;		

\$ Create two tables, one corresponding to LeftSeq and one to RightSeq. Loop through the
\$ items in each sequence. For each value of x, calculate the slope of the secant line
\$ determined by $P(a,f(a))$ and $Q(x,f(x))$.
 	
for x in LeftSeq do
    writeln x, (f(x) – f(a))/(x – a);
end for;
 	
for x in RightSeq do
    writeln x, (f(x) – f(a))/(x – a);
end for;		

b. Print copies of the two tables. Place them in your Activity Guide. Label the columns in the tables.

5

c. Based on your tables, draw some conclusions about the limiting behavior of the slopes of the secant lines determined by $P(a,f(a))$ and $Q(x,f(x))$ when $a = 1$.

Slope of secant lines determined by $P(1,4)$ $\rightarrow$ ☐ as $x \rightarrow 1^-$

and

Slope of secant lines determined by $P(1,4)$ $\rightarrow$ ☐ as $x \rightarrow 1^+$

Therefore,

Slope of secant lines determined by $P(1,4)$ $\rightarrow$ ☐ as $x \rightarrow 1$

d. Does the conclusion you reached using a numeric approach support the conclusion you reached in part 1 using a graphic approach? If not, look over your work again.

3. One more time. Use a symbolic approach to find the exact value of the slope of the tangent line to the $f(x) = x^2 + 3$ at $x = 1$.

a. Find a general expression for the slope of the secant line through $P(1,4)$ and $Q(x,f(x))$, where $f(x) = x^2 + 3$; that is, find an expression (in terms of x) for

$$\frac{f(x) - f(1)}{x - 1}$$

b. Calculate the actual value of the slope of the tangent line to the graph of $f(x) = x^2 + 3$ at $x = 1$ by finding the limit of the general expression for the slope of the secant lines (which you developed above in part a) as $x \rightarrow 1$.

c. This result should agree with the conjectures you made based on the graph in part 1 and your ISETL tables in part 2. Does it? If not, try to find where you went astray....

In the specific example considered in the last task, you analyzed the behavior of the slopes of the secant lines as they approached the tangent line to the graph at a particular point. You estimated the slope of the tangent line using graphic and numeric approaches, and then found the exact value by taking the appropriate limit. In the next task, you will generalize your observations and, in the process, discover a definition for the derivative.

The underlying idea is as follows: As long as the graph of f is smooth at $x = a$—that is, as long as f is continuous and its graph does not have a sharp peak or dip at $x = a$—you can:

- Construct two sequences, one approaching a from the left and one from the right.
- Find an expression for the slope of the secant line determined by $P(a, f(a))$ and the point on the graph corresponding to an item in the sequence.
- Find the slope of the tangent line by taking the limit of the slopes of the secant lines as the items in the sequence move closer and closer to a.

This is the same approach that you used in the last task to find the slope of the tangent line to the graph of $f(x) = x^2 + 3$ at $x = 1$. There will be one notational change, however. The items in the sequence that get closer and closer to a will be defined in terms of $a + h$, where $h \to 0$, instead of in terms of x where $x \to a$. This change in notation reflects the way LeftSeq and RightSeq are constructed.

Task 5-2: Discovering a Definition for the Derivative

1. Construct a general sequence that approaches $x = a$ from both sides.

a. Consider the sequence

```
RightSeq:= [a + 10**(−i): i in [1..5]];
```

The items in RightSeq approach $x = a$ from the right. Each item in RightSeq is formed by adding a small positive number h to a, that is, each item has the form

$$a + h, \quad \text{where in this case } h = 10^{-i}$$

Analyze the limiting behavior of $a + h$ as h approaches 0 from the right.

(1) Let h be a small positive number. Mark $a + h$ on the axis below.

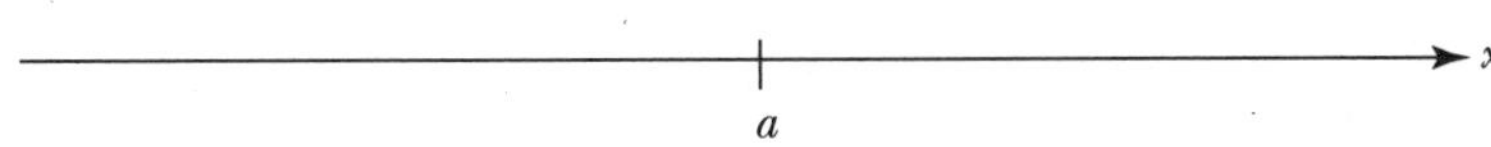

(2) Describe what happens to the value of $a + h$ as $h \to 0^+$. Support your description by drawing an arrow on the diagram in part (1).

b. Now look to the left of a. A sequence such as

```
LeftSeq := [a – 10**(–i): i in [1..5]];
```

approaches $x = a$ from the left, where the items in LeftSeq have the form

$$a + h, \quad \text{where } h = -(10^{-i})$$

Each item in the sequence is formed by adding to a negative number h close to 0. Analyze the limiting behavior of $a + h$ as h approaches 0 from the left.

(1) Let h be a negative number close to 0. Mark $a + h$ on the axis below.

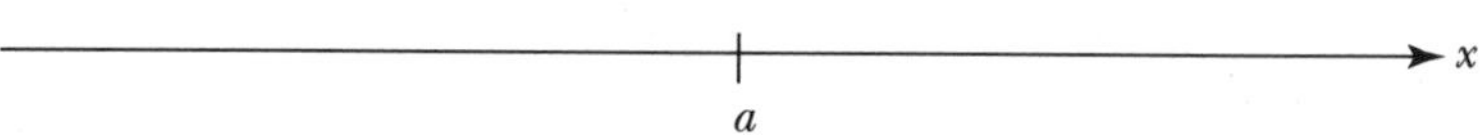

(2) Describe what happens to the value of $a + h$, as $h \to 0^-$. Support your description by drawing an arrow on the diagram in part (1).

c. Describe, in general, the limiting behavior of $a + h$ as $h \to 0$—that is, as h approaches 0 from both sides.

2. Find a general expression for the slope of the secant line determined by $x = a$ and a point close to a.

a. Consider $a + h$, which is near a but to the right of a.

(1) Sketch the secant line corresponding to a and $a + h$. Label the diagram given below as follows:

(a) Mark $a + h$ on the x-axis, where h is a small positive number.

(b) Mark $f(a)$ and $f(a + h)$ on the y-axis.

(c) Label the points $P(a,f(a))$ and $Q(a + h,f(a + h))$ on the graph of the function.

(d) Sketch the secant line determined by P and Q.

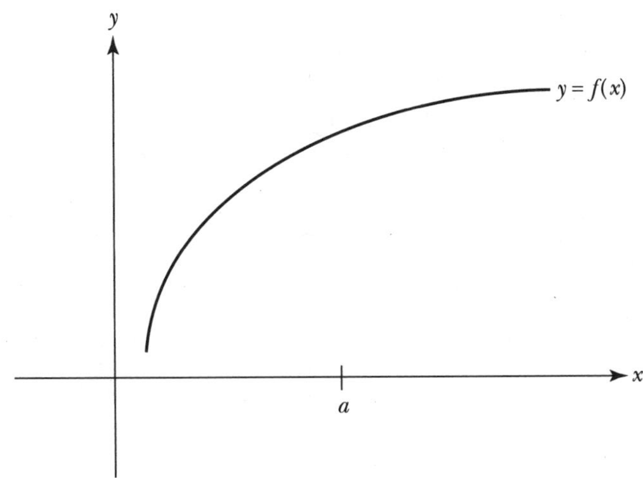

(2) Find an expression for the slope of the secant line in terms of the coordinates of P and Q—that is, in terms of a, $f(a)$, $a + h$, and $f(a + h)$. Simplify your expression.

(3) Describe the relationship between the values of the slopes of the secant lines as $h \to 0^+$ and the value of the slope of the tangent line at the point $P(a, f(a))$.

b. Do the same thing that you did above, but this time approach a from the left; that is, consider $a + h$, which is near a but to the left of a.

(1) Sketch the secant line corresponding to a and $a + h$. Label the diagram given below as follows:

(a) Mark $a + h$ on the x-axis, where h is a negative number close to 0.

(b) Mark $f(a)$ and $f(a + h)$ on the y-axis.

(c) Label the points $P(a, f(a))$ and $R(a + h, f(a + h))$ on the graph of the function.

5

(d) Sketch the secant line determined by P and R.

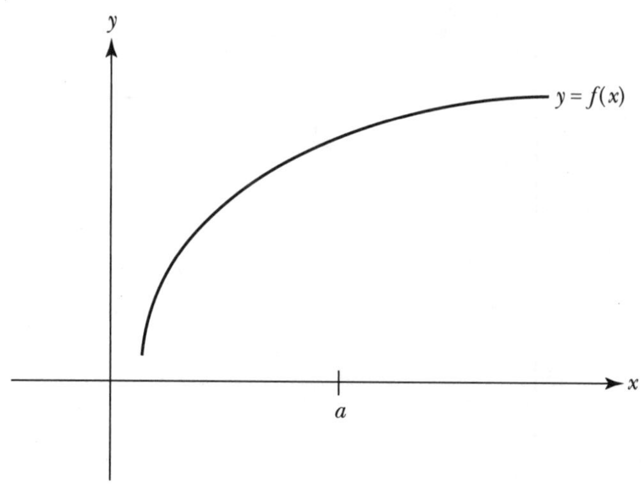

(2) Find the slope of the secant line in terms of the coordinates of P and R. Simplify your expression.

(3) Describe the relationship between the values of the slopes of the secant lines as $h \to 0^-$ and the value of the slope of the tangent line at the point $P(a, f(a))$.

3. Find the slope of the tangent line by taking the appropriate limit of the expression for the slope of the secant line determined by $x = a$ and $x = a + h$.

Congratulations: You have just developed the notation for what mathematicians call the *derivative,* one of the most important and fundamental concepts in calculus!

In particular, the derivative of a function $y = f(x)$ at $x = a$ is given by

$$\underbrace{\lim_{h \to 0} \underbrace{\frac{f(a+h) - f(a)}{h}}_{\substack{\text{slope of secant line deter-} \\ \text{mined by } P(a, f(a)) \text{ and} \\ Q(a+h, f(a+h))}}}_{\substack{\text{slope of tangent line to graph of} \\ f \text{ at } P(a, f(a))}}$$

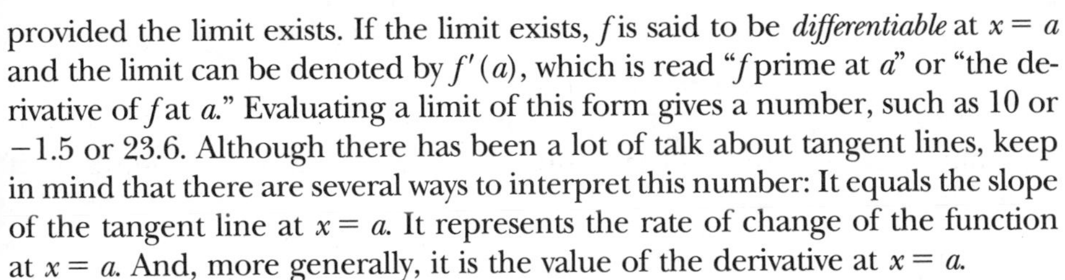

provided the limit exists. If the limit exists, f is said to be *differentiable* at $x = a$ and the limit can be denoted by $f'(a)$, which is read "f prime at a" or "the derivative of f at a." Evaluating a limit of this form gives a number, such as 10 or -1.5 or 23.6. Although there has been a lot of talk about tangent lines, keep in mind that there are several ways to interpret this number: It equals the slope of the tangent line at $x = a$. It represents the rate of change of the function at $x = a$. And, more generally, it is the value of the derivative at $x = a$.

The derivative takes an x-value in the domain of f and, if the limit exists, returns the value of the slope of the tangent line to the graph of f at that point or the value of the rate of change at f at that point. Consequently, the *derivative of a function is itself a function,* where the domain of the derivative is contained in the domain of the underlying function. If f is the name of the underlying function, then the derivative of f is denoted by f', and you can find the *general derivative* of f, $f'(x)$, by evaluating the limit

$$f'(x) = \lim_{h \to 0} \frac{f(x+h) - f(x)}{h}$$

If f is a polynomial, evaluating a limit of this form leads to an expression, such as $2x$ or $x^2 + 6$. You can use the expression representing f' to find the value of the derivative at a particular input, such as $x = 2$.

In general, to find $f'(x)$ using the definition of derivative, you do the following:

- Write down the *difference quotient*

$$\frac{f(x+h) - f(x)}{h}$$

- Simplify the difference quotient. Try to eliminate the h in the denominator so you can evaluate the limit by substituting zero for h.
- Take the limit as $h \to 0$.

In the next task, you will represent the derivatives of some functions by expressions and then use the expressions to analyze the functions.

Task 5-3: Representing the Derivative by Expressions

Find the derivative of a function using the definition. Use the derivative to calculate some tangent lines and analyze the behavior of the function. Before you begin, recall how to evaluate $f(x+h)$.

1. If $f(x) = -2x^2 + x - \frac{1}{2}$, you find $f(x+h)$ by evaluating f at $x + h$; that is, by substituting $x + h$ in for x.

$$\begin{aligned} f(x+h) &= -2(x+h)^2 + (x+h) - \tfrac{1}{2} \\ &= -2(x^2 + 2xh + h^2) + (x+h) - \tfrac{1}{2} \\ f(x+h) &= -2x^2 - 4xh - 2h^2 + x + h - \tfrac{1}{2} \end{aligned}$$

Evaluate each of the following functions at $x + h$. Simplify the result, if possible.

a. $f(x) = 5x - 1$

b. $f(x) = 2 - 4x - x^2$

c. $f(x) = -99$

d. $f(x) = \sqrt{x+4}$

e. $f(x) = \frac{1}{2}\sin(2x)$

f. $f(x) = 2^x$

g. $f(x) = \dfrac{x^2 - 1}{3x + 6}$

2. Consider $f(x) = x^2 + 3$, which you considered in Task 5-1.

a. Use the definition of derivative to find an expression for $f'(x)$.

$f'(x) = \lim\limits_{h \to 0} \dfrac{f(x+h) - f(x)}{h}$	Definition of $f'(x)$
$= \lim\limits_{h \to 0}$ ____________________	Evaluate $f(x+h)$ and $f(x)$
$= \lim\limits_{h \to 0}$ ____________________	Expand terms
$= \lim\limits_{h \to 0}$ ____________________	Simplify, factoring h out of the numerator
$= \lim\limits_{h \to 0}$	Cancel h's
$f'(x) =$	Take limit

b. Use the derivative to find the tangent line to the graph of f at the point $P(1,4)$.

(1) Find the slope of the tangent line to the graph of f when $x = 1$; that is, find $f'(1)$.

(2) Use the point-slope formula to find the equation of the tangent line to the graph of f when $x = 1$.

Hint: The tangent line touches the graph of f at P(1,4). You found its slope in part (1).

5

(3) Use your CAS to graph f and the tangent line to the graph of f when $x = 1$ on the same pair of axes, for $-1 \leq x \leq 4$. Place a copy of the graphs in the space below.

c. Use the derivative to analyze the behavior of f.

As you respond to the following queries, keep in mind that you are dealing with two functions: $f(x) = x^2 + 3$ and $f'(x) = 2x$. The expression representing f' can be used to calculate the derivative at any x-value where it exists. The value of the derivative provides information about the behavior of f. For instance, you can find the derivative when $x = 4$ by evaluating f' at $x = 4$:

$$f'(4) = 2 \cdot 4 = 8$$

This tells you that the slope of the tangent line to the graph of $f(x) = x^2 + 3$ at $P(4,f(4))$—that is, at $P(4,19)$—is 8. Moreover, because the slope of the tangent line to the graph of f is positive, you know that f is increasing at $x = 4$.

(1) Fill in the entries in the following table. In each case, indicate what the value of f' tells you about the behavior of the graph of f (increasing, decreasing, levels off).

x	f(x)	f'(x)	Graph of f Increasing, Decreasing, or Level at P(x, f(x))?
−3			
−1			
0			
2			
5			

(2) Use the expression for f' to find the set of all x-values where f is increasing—that is, where $f'(x) > 0$. Use interval notation to express your answer.

(3) Use the expression for f' to find the set of all x-values where f is decreasing—that is, where $f'(x) < 0$.

d. Find the (x, y)-coordinates of the point on the graph where the slope of the tangent line is 50.

e. Sketch the graphs of the functions f and f' on the same pair of axes. Label the graphs.

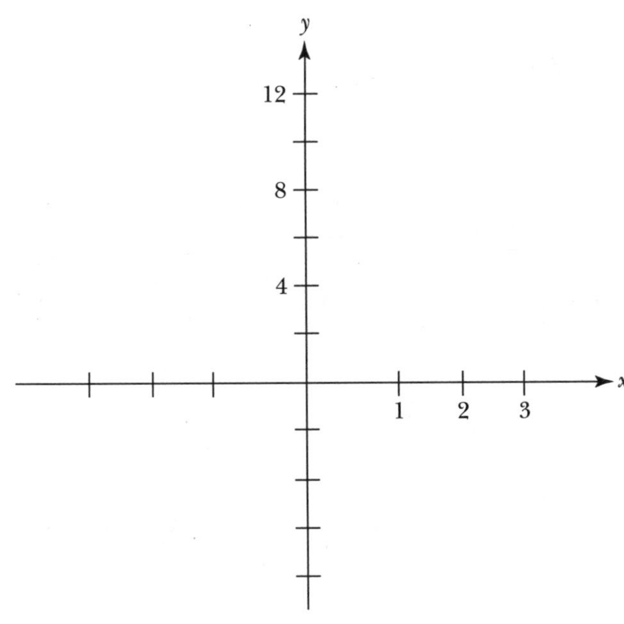

3. Consider $f(x) = 3x - 4$.

   a. Use the definition of derivative to find an expression for $f'(x)$.

   b. Use the expression for $f'(x)$ to find the rate of change of f at $x = 12$.

   c. Use the expression for $f'(x)$ to show that the graph of f is always increasing.

   d. The graph of f has no turning points. Explain how your expression for $f'(x)$ supports this conclusion.

   e. Graph the functions f and f' on the same pair of axes. Label the graphs

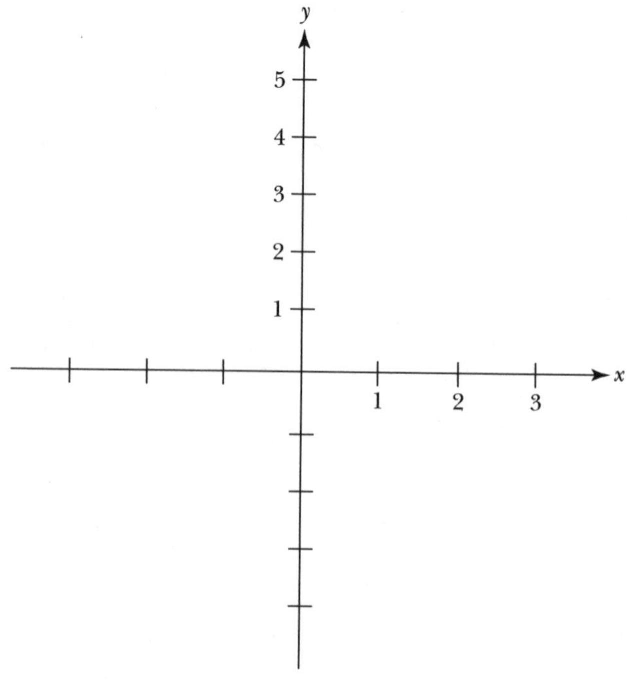

The derivative f' of a function f is a function. In the last task, you used the definition of derivative to represent f' by an expression. The question in the next task is: What is the domain of this new function? You know that the domain of f' is contained in the domain of f. But, are the two domains equal? Not necessarily. Just because a function is defined at $x = a$ does not guarantee that the tangent line at $P(a, f(a))$ exists. It is possible that $f(a)$ exists and $f'(a)$ does not, in which case the limit

$$\lim_{h \to 0} \frac{f(a+h) - f(a)}{h}$$

fails to exist. Thinking back to your study of limits, you know that:

- A limit does not exist if its left- and the right-hand limits exist, but have different values. Consequently, $f'(a)$ does not exist if the left and right limits, which are called the *left-* and *right-hand derivatives* exist, but have different values; that is, if

$$\lim_{h \to 0^-} \frac{f(a+h) - f(a)}{h} \neq \lim_{h \to 0^+} \frac{f(a+h) - f(a)}{h}$$

- A limit does not exist if its output values explode as the input values get closer and closer to the designated number. Consequently, $f'(a)$ does not exist if

$$\frac{f(a+h) - f(a)}{h}$$

blows up as h approaches 0 from the left and/or the right; that is, if

$$\lim_{h \to 0^-} \frac{f(a+h) - f(a)}{h} = \pm\infty \qquad \text{and/or} \qquad \lim_{h \to 0^+} \frac{f(a+h) - f(a)}{h} = \pm\infty$$

In the next task, you will examine what goes wrong when you try to evaluate the limit at a place where the derivative does not exist. As you do the task, keep two things in mind: First, the following are equivalent ways of saying that $x = a$ is not in the domain of f':

- f is not differentiable at $x = a$.
- $f'(a)$ does not exist.
- f' is undefined at $x = a$.

Second, as a line becomes more and more vertical, the slope of the line approaches $+\infty$ or $-\infty$. Consequently, the slope of a vertical line is undefined.

Task 5-4: Inspecting the Domain of the Derivative

The domain of f' is contained in the domain of f. What points are left out?

1. If the graph of a function f has a sharp "peak" or "dip" at $x = a$, then $f'(a)$ does not exist. Examine why this is true.

Actually, this is not a new idea. Remember the dude with the pointy hair in Unit 1? There were no tangent lines on the tips of his hair...

a. Consider the graph of $f(x) = |x|$, which has a sharp dip at $x = 0$.

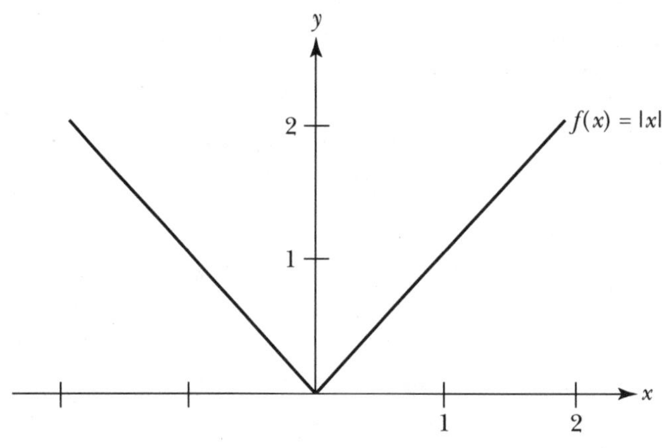

(1) Use a graphic approach to find the left-hand derivative of f at $x = 0$.

The *left-hand derivative* of f at $x = 0$ is the limit of the slopes of the secant lines determined by $P(0,f(0))$ and Q, where Q approaches P from the left. Consequently, one way to determine the left-hand derivative of f at $x = 0$ is to find the slope of the secant line determined by $P(0,f(0))$ and say $Q(-1,f(-1))$. Next, find the slope of the secant line for another value of Q which is closer to P, but still to its right, for instance, $Q(-0.5,f(-0.5))$. Find the value of the left-hand derivative by generalizing your observations. Record your conclusion below.

The left-hand derivative of f at $x = 0$ is __________ .

(2) Use a graphic approach to find the right-hand derivative of f at $x = 0$.

The *right-hand derivative* of f at $x = 0$ is the limit of the slopes of the secant lines determined by $P(0, f(0))$ and Q, where Q approaches P from the right. Repeat the process you used to find the left-hand derivative in part (1), but this time consider a couple of values of Q which are close to $P(0, f(0))$ but to the right of P. Record your conclusion below.

The right-hand derivative of f at $x = 0$ is _________ .

(3) Explain why $f(x) = |x|$ is not differentiable at $x = 0$.

b. Generalize your observations.

(1) Sketch an arbitrary function f that is defined at $x = a$ and has a sharp peak or dip at $x = a$.

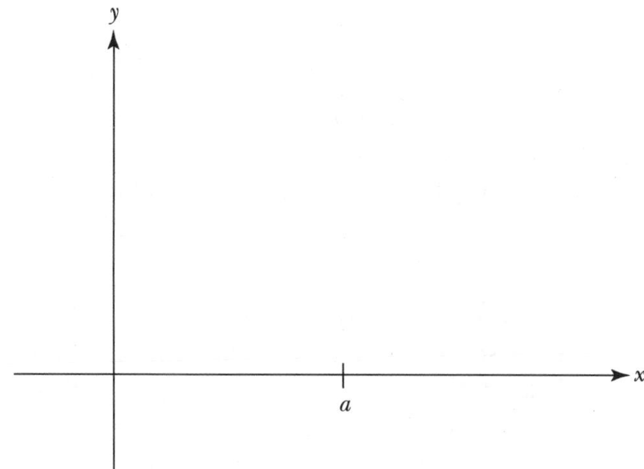

(2) Explain why f is not differentiable at $x = a$.

2. If the graph of a function f has a jump discontinuity at $x = a$, then $f'(a)$ does not exist. Examine why this is true.

a. Consider the function

$$f(x) = \begin{cases} x + 1, & \text{if } x \leq 2 \\ 5, & \text{if } x > 2 \end{cases}$$

The graph of f has a jump discontinuity at $x = 2$.

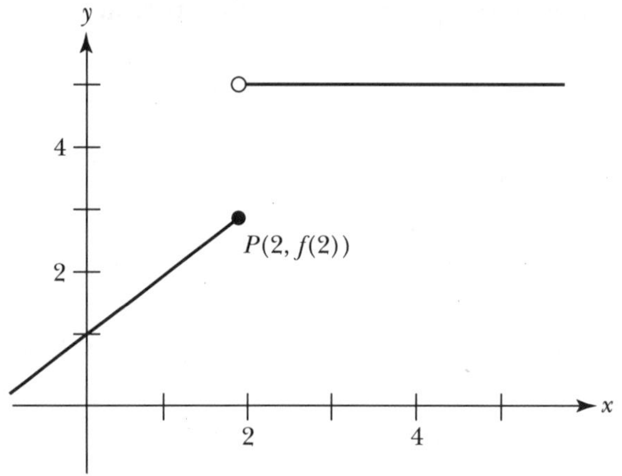

(1) Find the value of the left-hand derivative of f at $x = 2$ by calculating the slope of several secant lines determined by $P(2,f(2))$ and some points on the graph of f to the left of P.

Left-hand derivative of f at $x = 2$ is ________.

(2) Explain why the right-hand derivative of f at $x = 2$ does not exist by sketching several secant lines determined by $P(2,f(2))$ and some points on the graph of f to the right of P.

(3) Explain why f is not differentiable at $x = 2$.

b. Generalize your observations.

(1) Sketch an arbitrary function f that has a jump discontinuity at $x = a$. Label the point $P(a, f(a))$ on your graph.

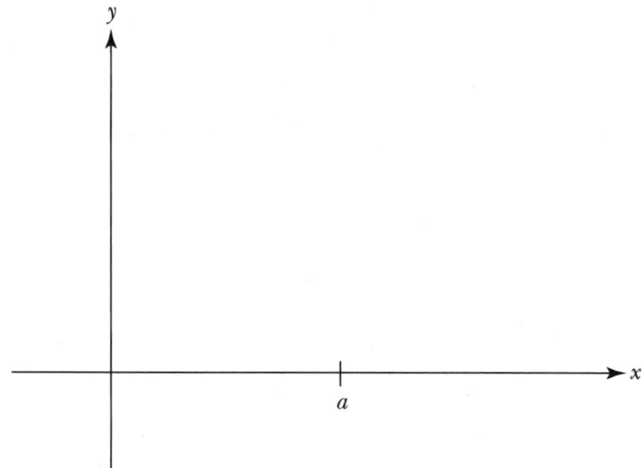

(2) Explain why f is not differentiable at $x = a$. In the case of your function, which one-sided derivative does not exist?

3. If the graph of a function f has a vertical tangent at $x = a$, then $f'(a)$ does not exist. Examine why this is true.

Consider the graph of the following function which has a vertical tangent at $x = a$.

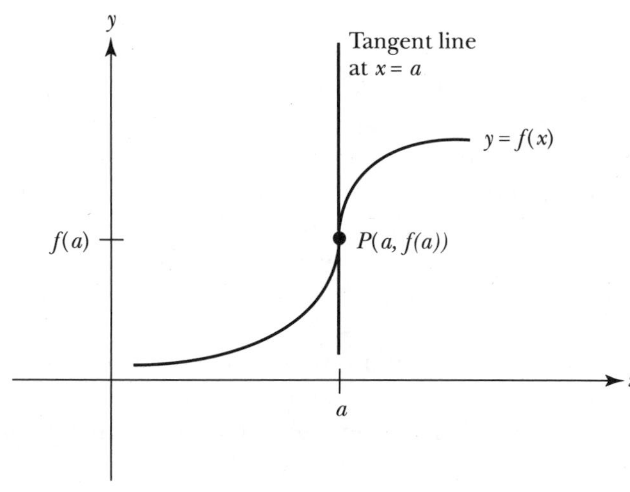

5

a. Explain why the left-hand derivative of f at $x = a$ does not exist. Sketch the secant lines determined by $P(a, f(a))$ and several points on the graph of the function to the left of P. Describe what happens to the value of the slope of the secant line as the points get closer and closer to P.

b. Using a similar approach as in part a, explain why the right-hand derivative of f at $x = a$ does not exist.

c. Explain why f is not differentiable at $x = a$.

In Unit 1, you used the motion detector to create distance versus time graphs, and you analyzed the behavior of the tangent line as it traveled along the graphs. You discovered that the value and sign of the slope of the tangent line provide a lot of information about the shape and behavior of the graph, including the following facts:

- The value of the slope of the tangent line corresponds to your velocity or instantaneous rate of change at a given time.
- The magnitude of the slope of the tangent line corresponds to the speed you are moving at a particular time.
- The sign of the slope indicates whether the function modeling your motion rises or falls from left to right. The sign is positive when the graph is increasing and negative when it is decreasing.
- If the value of the slope is zero, the graph of the function has a horizontal tangent.
- The change in the sign of the slope as the tangent line travels along the graph from left to right indicates the location of the function's local extrema. If the sign of the slope changes from positive to zero to negative, the function has a local maximum. If it changes from negative to zero to positive the function has a local minimum.
- If the graph lies above the tangent line, the function is concave up. If it lies below, it is concave down.

In the next task, you will reexamine these ideas and express them in terms of the derivative. You will also investigate how the graph of the derivative is related to the graph of the given function. In particular, based on behavior of f, you will determine the following:

- The places where f' is undefined
- The places where the graph of f' intersects the horizontal axis
- The intervals where the graph of f' lies above the horizontal axis and the intervals where the graph lies below

As you explore the relationship between f and f', you will develop the statement for the First Derivative Test.

Task 5-5: Investigating the Relationship Between a Function and Its Derivative

1. Summarize what you know about the relationship between a function and the tangent line to its graph. Express these ideas in terms of the derivative and describe their impact on the graph of the derivative.

a. Consider the cases where a function f is increasing, decreasing, or has a horizontal tangent at $x = a$.

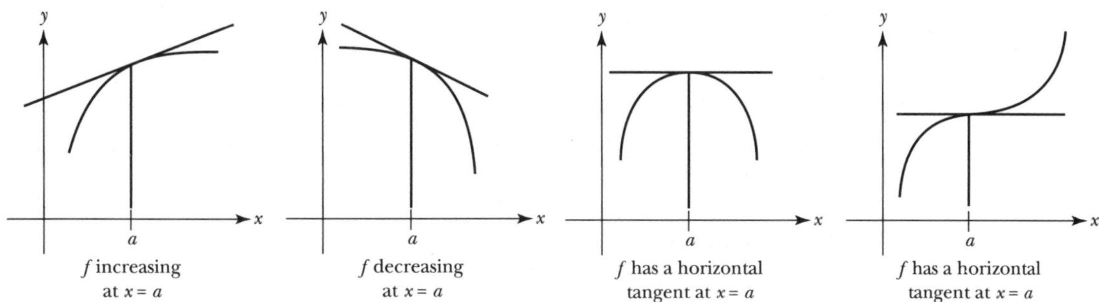

Summarize what's happening in each of these situations by filling in the following chart. The first row has been completed for you. Look it over carefully before completing the other rows.

Description of f at $x = a$	Shape of Graph of f near $x = a$	Sign of Slope of TL at $x = a$. i.e., Sign of $f'(a)$	Location of Graph of f' at $x = a$
f increasing	Rises from left to right	+	Above horizontal axis
f decreasing			
f has a horizontal tangent			

Note: TL stands for "tangent line."

b. Consider the cases where a function f has a local minimum, a local maximum, or a place where the graph levels off at $x = a$.

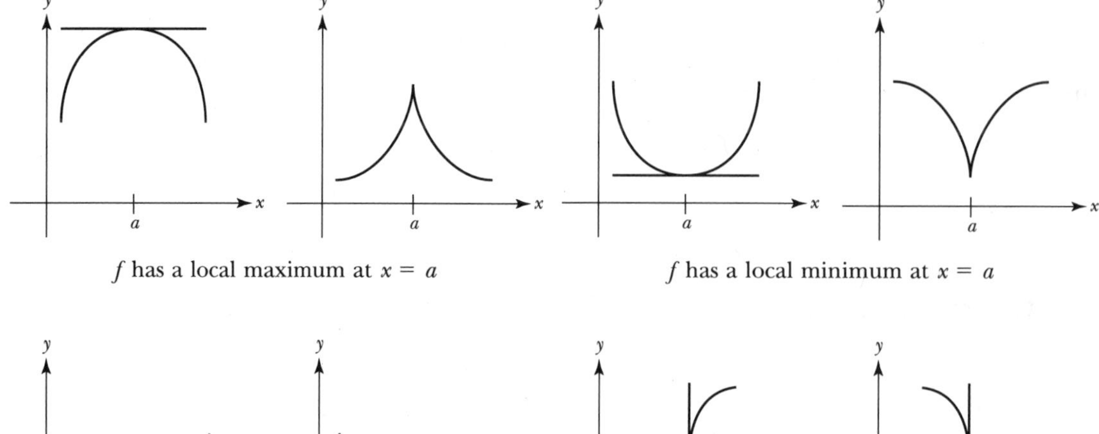

f has a local maximum at $x = a$ f has a local minimum at $x = a$

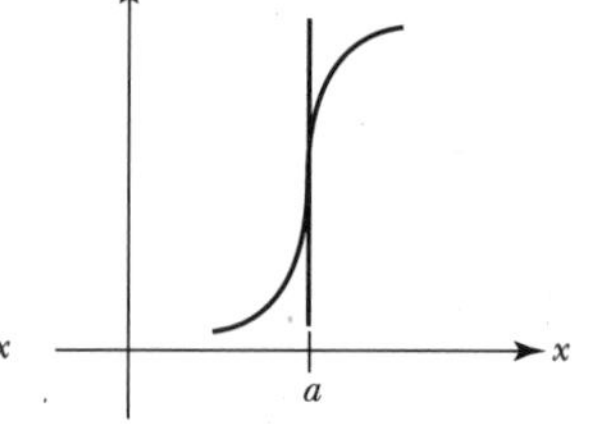

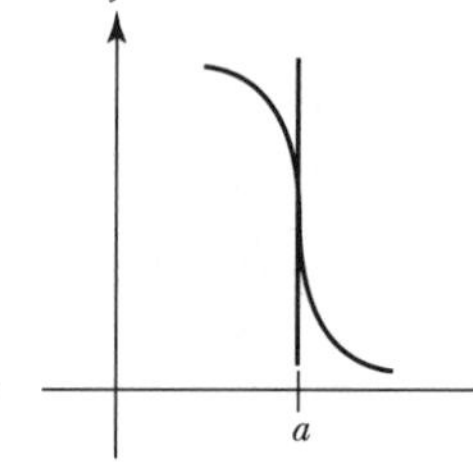

f levels off at $x = a$ f has a vertical tangent at $x = a$

Summarize what's happening in each of these situations by filling in the following chart. The first row has been completed for you. As usual, look at it carefully before completing the other rows.

Note: DNE stands for "does not exist."

Description of f at $x = a$	Shape of Graph of f near $x = a$	Change in Sign of Slope of TL near $x = a$. i.e., Change in Sign of $f'(x)$	Behavior of Graph of f' near $x = a$
f has a local maximum	Increases to left of $x = a$ and decreases to right.	Changes from + to 0 at $x = a$ to −, or from + to DNE at $x = a$ to −.	Falls from left to right. Crosses axis at $x = a$. Either 0 or undefined at $x = a$.
f has a local minimum			
f levels off			
f has a vertical tangent			

2. Suppose f is a an arbitrary function. How are f and f' related? What does f' tell you about the graph of f? What does f tell you about the behavior of f'?

Consider the graph of f.

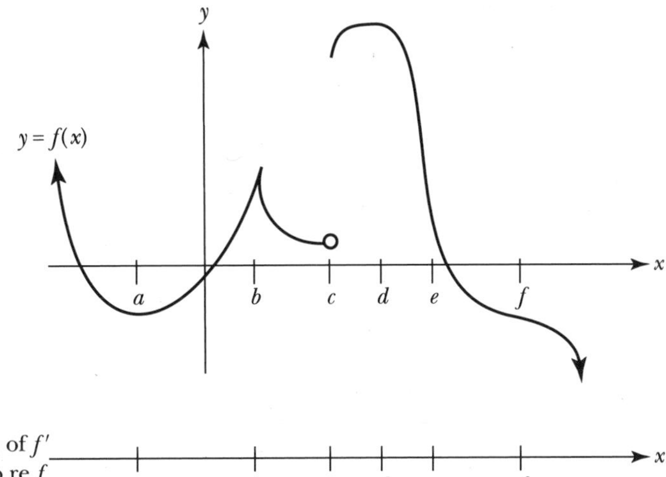

a. Fill in the Info re f (information regarding f) on the chart given below the graph.

(1) Label the intervals where f is increasing (inc) and decreasing (dec).

(2) Mark the x-values where f has a local maximum (LM), has a local minimum (lm), and levels off (LO).

b. Investigate the relationship between the shape of the graph of f and the sign of f' as you fill in the sign chart for f'.

(1) Find all the x-values where $f'(x)$ does not exist. Write DNE above each value on the sign chart for f'. Note that these are the x-values where f' is undefined.

(2) Find all the x-values such that $f'(x) = 0$, that is, where the slope of the tangent line to the graph of f is 0. Mark the values with a 0 on the sign chart for f'. Note that these are the x-values where the graph of f' intersects the horizontal axis.

(3) Identify all the intervals where $f'(x) < 0$, that is, where the slope of the tangent line to the graph of f is negative since f is decreasing. Place a string of $-$'s above each interval on the sign chart for f'. Note that these are the intervals where the graph of f' lies below the horizontal axis.

(4) Identify all the intervals where $f'(x) > 0$, that is, where the slope of the tangent line to the graph of f is positive since f is increasing. Place a string of +'s above each interval on the sign chart for f'. Note that these are the intervals where the graph of f' lies above the horizontal axis.

(5) Summarize what you know about the graph of f'.

What if you do not know what the graph of f looks like, but you do know the sign chart for f'? How can you locate the local extrema by examining the sign chart for f'? The First Derivative Test, which follows directly from the observations you made in the last task, provides you with a way of identifying local extrema.

The First Derivative Test

Suppose $f'(c) = 0$ or $f'(c)$ does not exist. Suppose I is an open interval containing c, where f is a continuous and differentiable everywhere on I, except possibly at $x = c$.

i. If $f'(x) > 0$ for all x immediately to the left of c and $f'(x) < 0$ for all x immediately to the right of c—that is, if f is increasing to the left of c and decreasing to the right of c—then f has a local maximum at $x = c$.

ii. If $f'(x) < 0$ for all x immediately to the left of c and $f'(x) > 0$ for all x immediately to the right of c—that is, if f is decreasing to the left of c and increasing to the right of c—then f has a local minimum at $x = c$.

The sign of the derivative of a function gives you a lot of information about the shape of the graph of its underlying function. If the derivative is positive over some interval, you know that the function is increasing over that interval. If the derivative is negative over some interval, the function is decreasing over that interval. If the derivative equals 0, the function has a horizontal tangent—which is either a local extremum or a leveling off point.

In addition to indicating the sign of the derivative on the sign chart, you can also indicate values where the derivative does not exist. This, too, provides you with information about the behavior of the graph of the underlying function. In particular, if the function is continuous at the point where the derivative does not exist, the graph of the function may have a cusp or a vertical tangent. On the other hand, if the function is not continuous at the point where the derivative does not exist, the graph of the function may have a jump.

In the following activities, you will interpret some sign charts for derivatives and see what they tell you about the shapes of the graphs of the underlying functions.

Task 5-6: Gleaning Information About the Graph of a Function from Its Derivative

1. Consider the following sign chart:

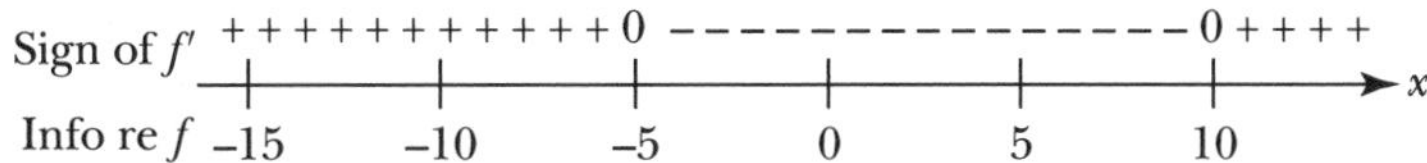

Assume f is defined for all real numbers. In addition, assume that the only zeros of f' are at $x = -5$ and $x = 10$. Use the sign chart for f' to answer the following questions about the shape of the graph of f.

a. Determine the intervals where f is increasing (inc). Record this information regarding f on the sign chart given above.

b. Determine the intervals where f is decreasing (dec). Record this information regarding f on the sign chart given above.

c. Find all the x-values where f has a local maximum (LM) or a local minimum (lm) and where the graph of f levels off (LO). Record this information regarding f on the sign chart given above.

d. Sketch a graph of a function f whose derivative has the given sign chart.

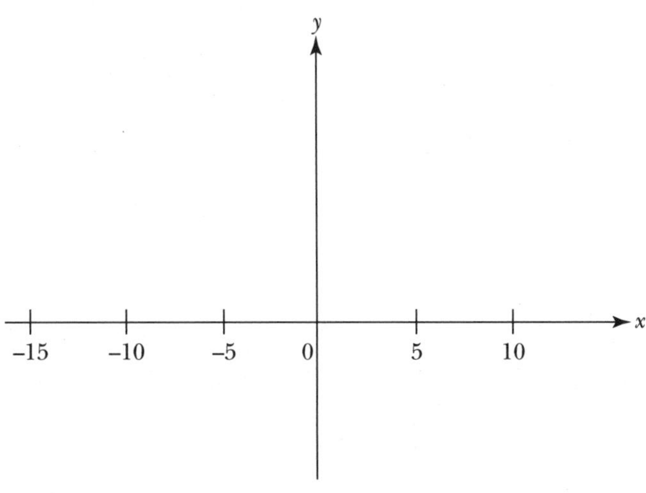

5

2. Consider the following graph of the derivative f':

Graph of f'

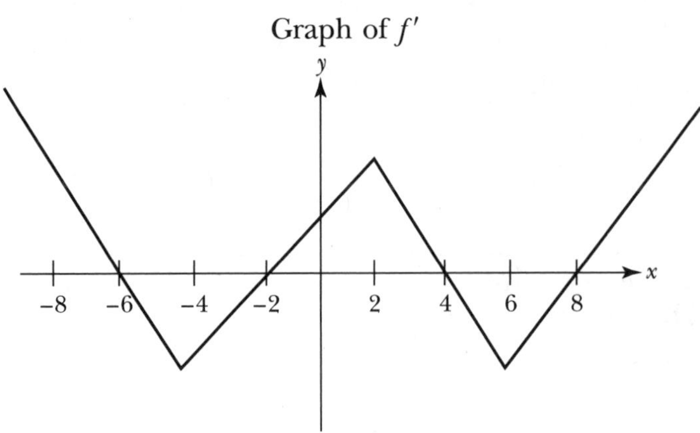

a. Find the associated sign chart for f'.

Sign of f'

Info re f −8 −6 −4 −2 0 2 4 6 8 x

b. Describe the shape of the graph of f by filling in information regarding f; that is, indicate the intervals where f is increasing and decreasing, and give the location of all the extrema of f.

c. Suppose the graph of f lies above the horizontal axis for $-8 \le x \le 8$—that is, $f(x) > 0$ for $-8 \le x \le 8$. Sketch a graph of a function f, where the graph of f' is given above.

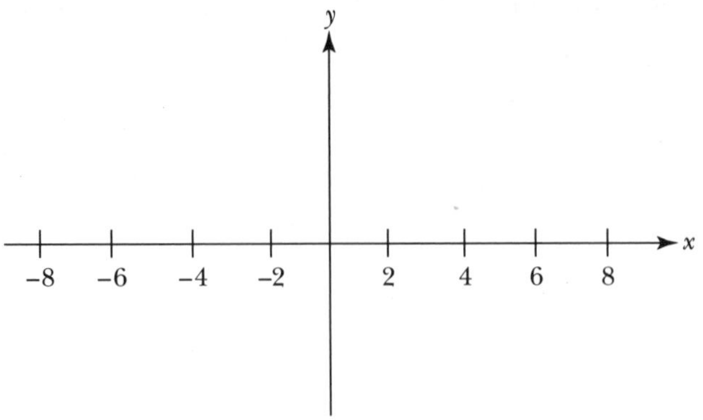

3. On the axes given below, sketch the general shape of the graph of a function f which satisfies the following conditions:

- f defined for all real numbers

- $f'(x) = x^2$
- $f(0) = 3$

Note: First, find the sign chart for the derivative f′ and fill in the "info re f."

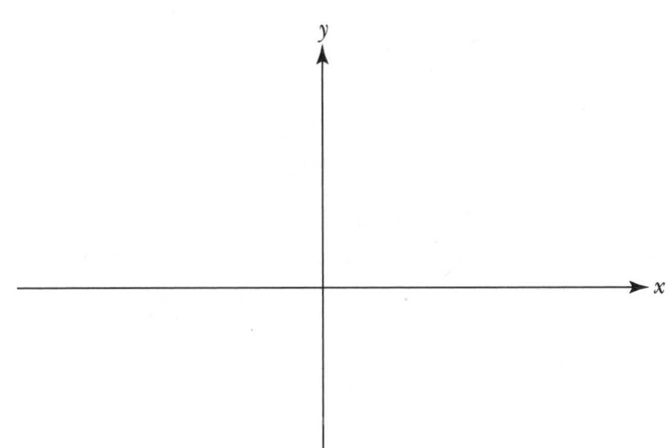

Unit 5 Homework After Section 1

- Complete the tasks in Section One in the Activity Guide. Be prepared to discuss them in class.
- Calculate some derivatives using the definition in HW5.1. Use the derivative to analyze the function. Compare the graphs of a function and its derivative.

HW5.1 Recall that, by definition,

$$f'(x) = \lim_{x \to 0} \frac{f(x + h) - f(x)}{h}$$

provided the limit exists.

1. For each of the following functions:

i. Use the definition to find its general derivative.

ii. Sketch the graphs of f and f' on the same pair of axes. Label the graphs.

a. $f(x) = 6$

b. $f(x) = 6.5x - 10$

c. $f(x) = x^3$. *Note:* $f(x + h) = (x + h)^3 = x^3 + 3x^2h + 3xh^2 + h^3$

5

2. Suppose $f(x) = x^2 - 2x + 5$.

   a. Use the definition to find $f'(x)$.

   b. Sketch the graphs of f and f' on the same pair of axes. Label the graphs.

   c. Use $f'(x)$ to find the equation of the tangent line at $P(2, f(2))$.

   d. Use the expression for $f'(x)$ to find the (x,y)-coordinates of the point on the graph of f where the tangent line is horizontal to the graph.

   e. Use the expression for $f'(x)$ to find the intervals where the function is increasing and the intervals where it is decreasing.

3. Use the definition of derivative to show that each of the following statements is true. Support the conclusion of the statement by graphing a typical f and its derivative f' on the same pair of axes. Label the graphs.

   a. The derivative of a constant function is always 0; that is, if $f(x) = c$, where c is a fixed constant, then $f'(x) = 0$ for all x.

   b. The derivative of a linear function is the slope its graph; that is, if $f(x) = mx + b$, where $m \neq 0$, then $f'(x) = m$ for all x.

- Use the derivative to analyze the shape of an underlying function in HW5.2.

HW5.2 Sketch the graph of a function that satisfies the given condition.

1. Sketch the graph of a function f satisfying the following conditions:

   - f is defined for all real numbers.
   - The sign chart of the derivative of f is

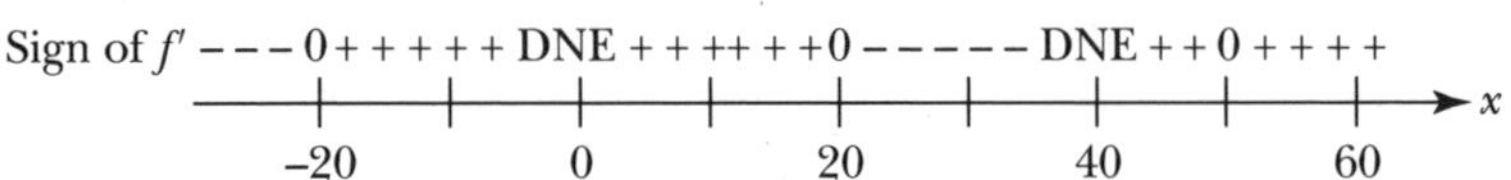

   where DNE indicates that the derivative does not exist.

2. Sketch the graph of a function s satisfying the following conditions:

   - s is continuous for all real numbers.
   - s is differentiable everywhere, except at $t = 6$—that is $s'(6)$ DNE.
   - $s'(t) \leq 0$ for all t, except $t = 6$.
   - $s'(3) = s'(9) = 0$.
   - $s(6) = 0$.

3. Sketch the graph of a function h satisfying the following conditions:

- h is continuous for all real numbers.
- $h'(x) = x^2 - x$.
- The graph of h passes through the point $P(\frac{1}{2}, -4)$.

4. Sketch the graph of a function f satisfying the following conditions:

- f is continuous for all real numbers.
- $f(x) \leq 0$, for all x.
- $f(10) = 0$.
- The graph of f' has the following shape:

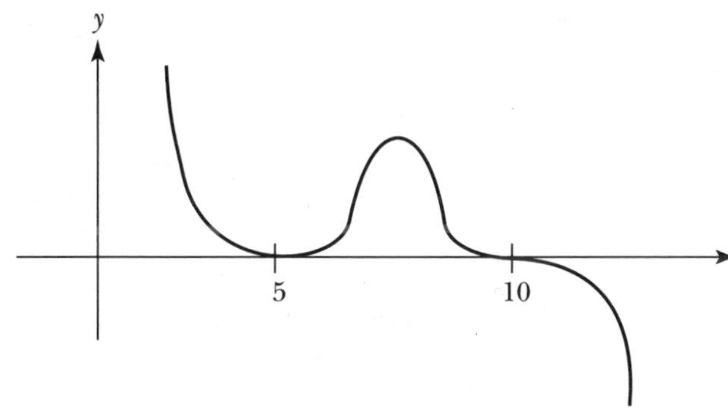

- Compare the graphs of a function and its derivative in HW5.3.

HW5.3 Each of the following graphs displays the graph of a function f and its derivative f'. Indicate which graph is which. Justify your choices.

1.

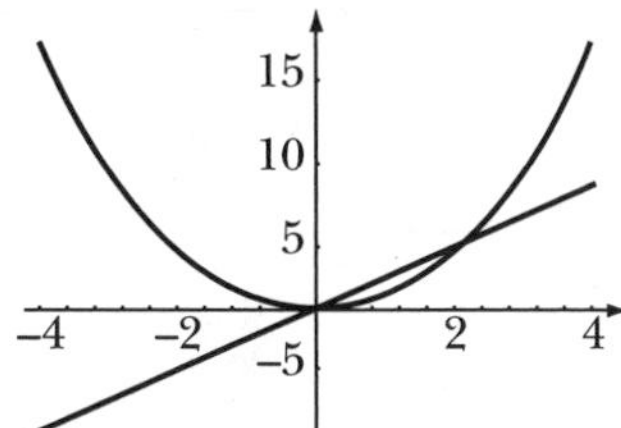

2.

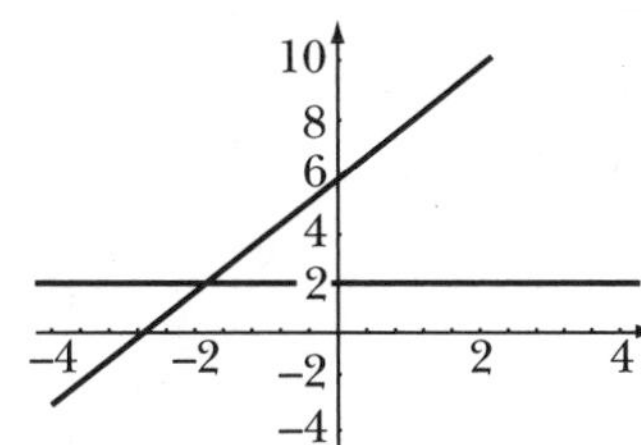

3.

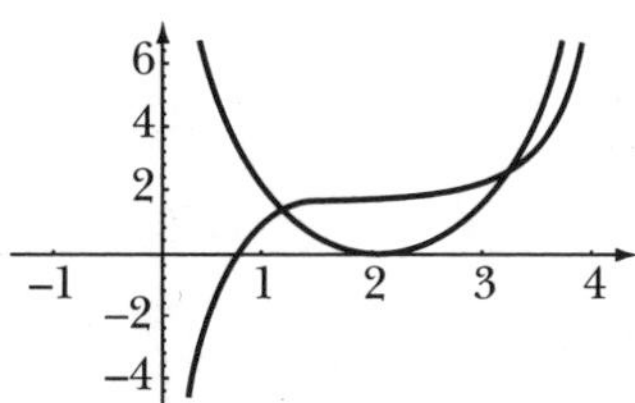

4.

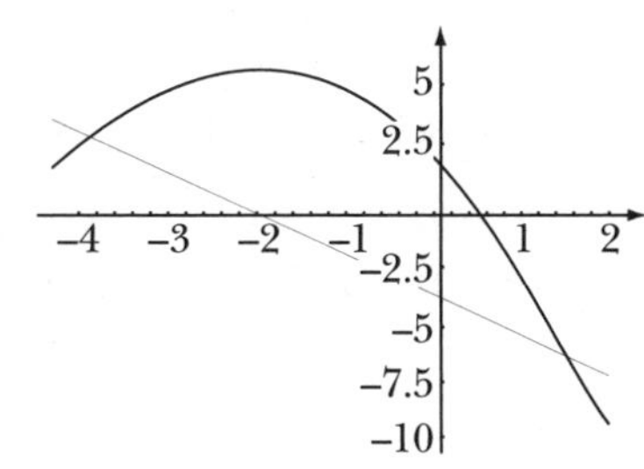

5

5.

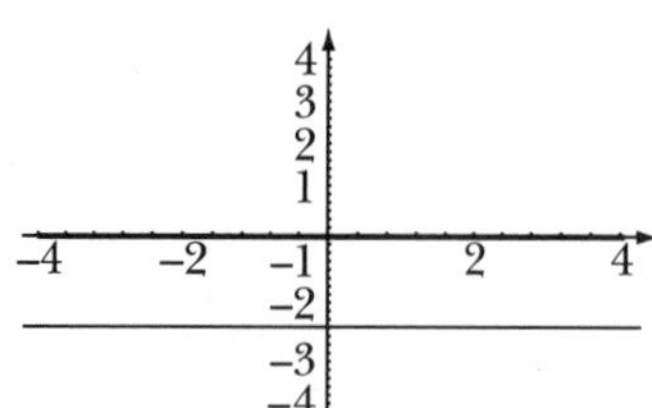

6.

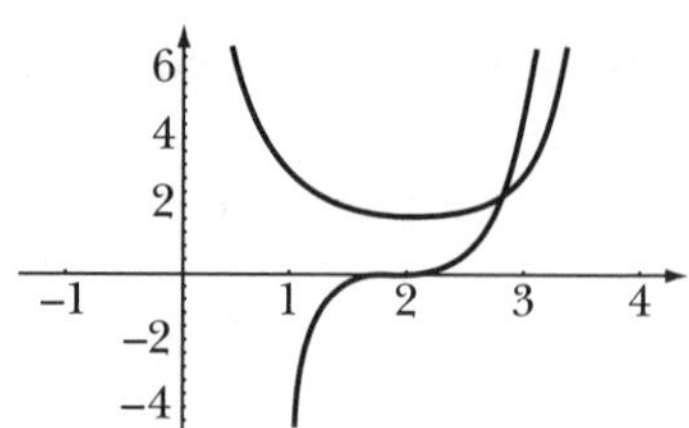

7.

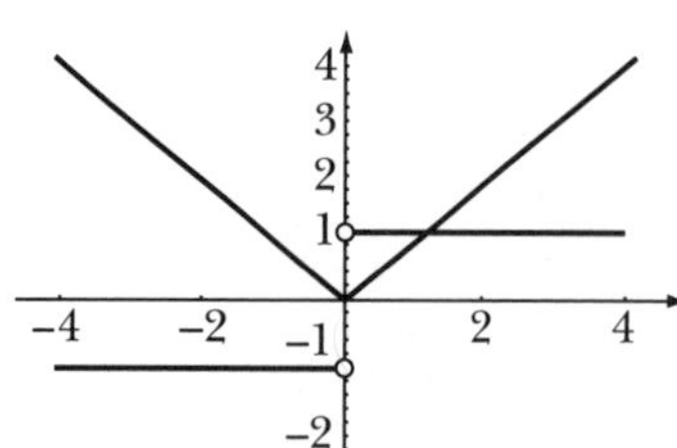

8.

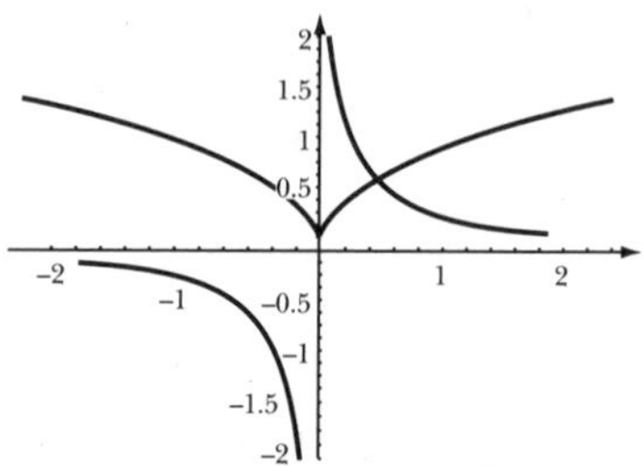

- Many different functions have the same derivative. Make a first pass at determining the relationship between two functions that have the same derivative in HW5.4.

HW5.4 Keep in mind that the value of f' can be determined by the value of the slope of the tangent line as it travels along the graph of f.

1. Suppose the graph of f' is given below:

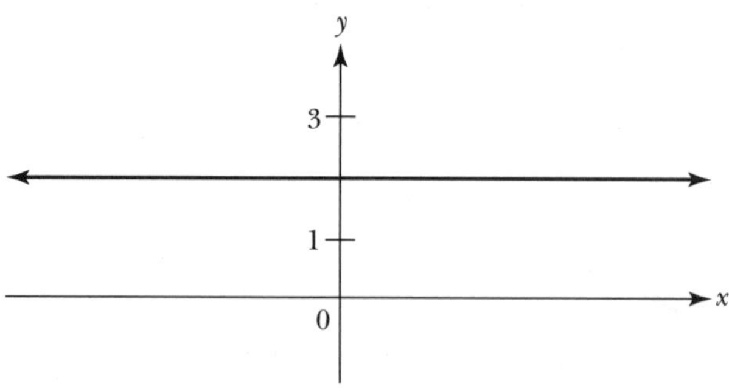

a. Based on the shape of the graph of f', explain why the graph of f is a line.

b. Give the slope of the graph of f.

c. Assume f passes through the point $P(0,0)$.

(1) Represent f by a graph. Label your axes.
(2) Represent f by an expression.

d. Instead of assuming the graph of f passes through the point $P(0,0)$, assume the graph of f passes through the point $P(0,-3)$.

(1) Represent f by a graph. Label your axes.
(2) Represent f by an expression.

2. Suppose the graph of f' is given below:

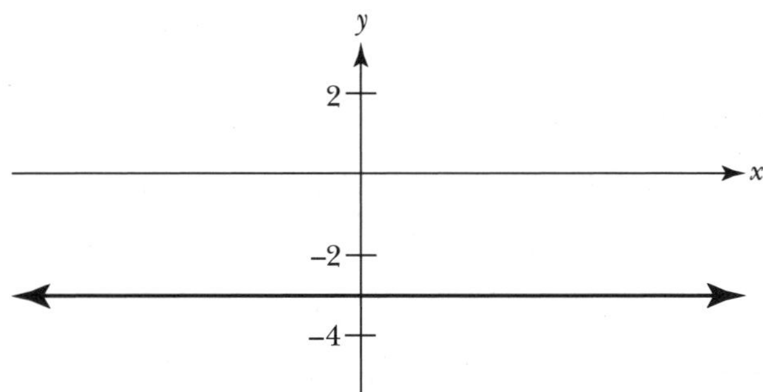

a. Assume $f(0) = 2$. Represent f by a graph and an expression.

b. Assume $f(2) = 0$. Represent f by a graph and an expression.

3. What's going on here? In each instance, there are at least two functions that have the same derivative. Describe the relationship between the graphs of linear functions that have the same derivative.

- The value of $f'(a)$ equals the slope of the tangent line to the graph of f at $x = a$. It also gives the rate of change of f at $x = a$. With this interpretation, instead of finding the limit of the slopes of secant lines, you think about taking the limit of average rates of change. Calculate some average rates of change in HW5.5.

HW5.5 Suppose b is a point near a in the domain of f. Then

average rate of change of f between $x = a$ and $x = b$

= slope of secant line determined by $P(a,f(a))$ and $Q(b,f(b))$

$$= \frac{f(a) - f(b)}{a - b}$$

$$= \frac{\text{change in output values}}{\text{change in input values}}$$

5

whereas

$$\begin{aligned}(\textit{instantaneous})\ \textit{rate of change} \text{ of } f \text{ at } x = a \\ &= \text{slope of the tangent line to the graph of } f \text{ at } P(a, f(a)) \\ &= \lim_{b \to a} \frac{f(a) - f(b)}{a - b} \\ &= f'(a)\end{aligned}$$

Calculate some average rates of change.

1. Fruit flies multiply fast. The following graph of the function N shows the growth of a fruit fly population during a 5-hour period.

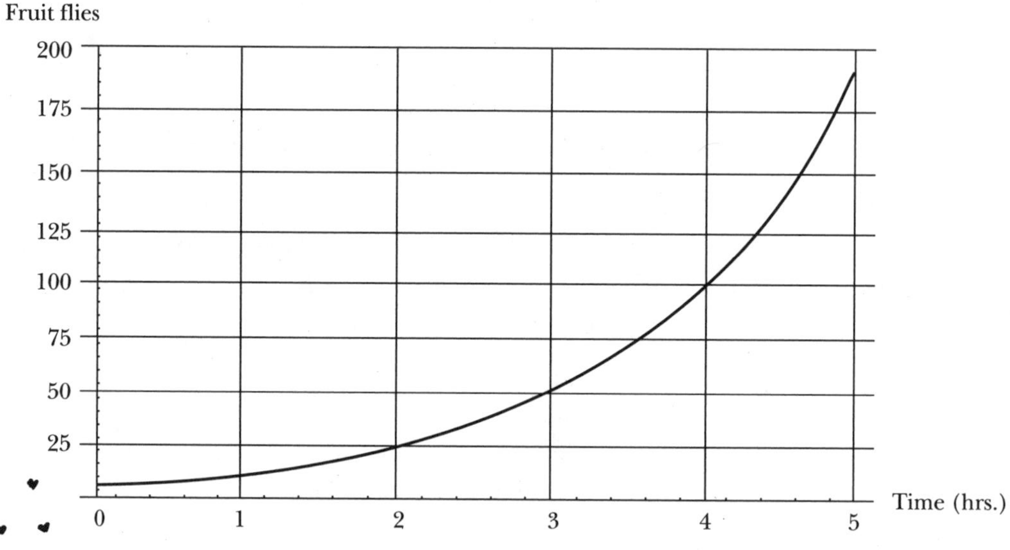

a. Approximate the numerical value for the average rate in the fruit fly population over the following time intervals:

(1) Between 3 and 5 hours
(2) Between 3 and 4 hours
(3) Between 1 hour, 30 minutes and 3 hours

b. Give a graphic representation of each of the numerical values which you found in part a by drawing the associated secant line on the graph of N. Label the lines (1), (2), and (3) to indicate which time interval the line represents.

2. The following table shows the growth chart for Mark:

Age (yr.)	0	2	4	6	8	10	12	14	16
Height (in.)	19	23	28	36	46	55	66	70	72

a. Graph Mark's growth (either by hand or using your CAS).

b. Describe Mark's growth rate during the 16-year period.

c. Find Mark's average rate of growth in inches per year from the time he was born until he was 4 years old.

d. Find Mark's average rate of growth in inches per year from age 12 to age 16.

e. Find the 2-year period during which time Mark had his largest average rate of growth.

f. Find the four year period during which time he had his slowest average rate of growth.

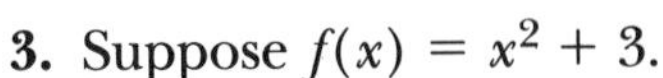

3. Suppose $f(x) = x^2 + 3$.

a. Find the average rate of change of f:

(1) Between $x = 1$ and $x = -1$
(2) Between $x = 1$ and $x = 0$
(3) Between $x = 1$ and $x = 2$
(4) Between $x = 1$ and $x = 3$

b. Graph the function f (either by hand or using your CAS) for $-1 \leq x \leq 3$.

c. On your graph, carefully sketch the secant lines determined by the pairs of input values given in part a. Label each secant line on your graph and indicate its slope.

- In the case of the distance function, the derivative gives the velocity of an object at any given time. Apply this interpretation of derivative as you do HW5.6.

HW5.6 Time is running out. Suppose you have 15 seconds to get to your destination 30 meters away. You start off at an acceptable pace, but because you are loaded down with books, you keep slowing down. The graph below shows your velocity function for the next 15 seconds.

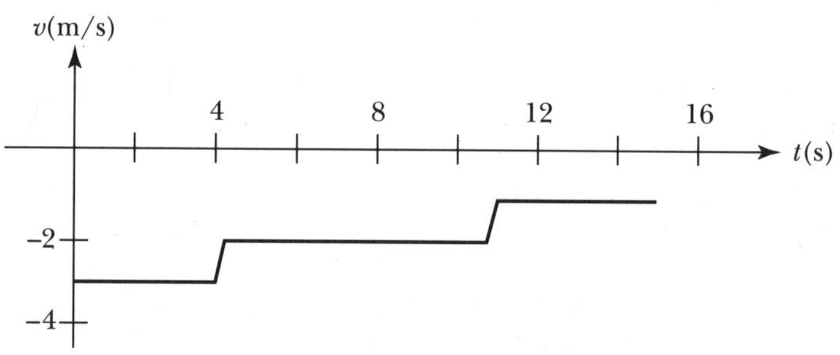

1. Give a verbal description of your movement. For instance, you might say, "During the first 4 seconds my velocity is -3 meters/second. Therefore, during the first 4 seconds, I am decreasing my distance from my destination, moving at a rate of 3 meters each second," and so on.

2. Approximate the total distance you move during the 15-second time period.

3. Recalling that initially you are 30 meters from your destination, sketch your underlying distance function.

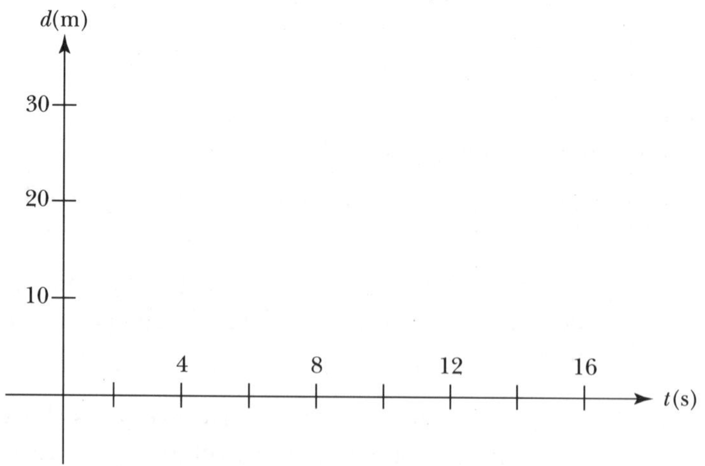

- Think about the meaning of the Mean Value Theorem for Derivatives in HW5.7.

HW5.7 According to the *Mean Value Theorem for Derivatives*:

If $y = f(x)$ is continuous on the closed interval $[a,b]$ and is differentiable at every point in the open interval (a,b), then there is at least one point c between a and b at which

$$\frac{f(b) - f(a)}{b - a} = f'(c)$$

1. Explain why the Mean Value Theorem for Derivatives makes sense. Develop your explanation as follows:

a. Sketch an appropriate diagram.

(1) Sketch a pair of axes, and label a and b on the horizontal axis where $a < b$.
(2) Sketch the graph of a "smooth" function f over $[a, b]$—that is, a function which is continuous on the closed interval $[a, b]$ and is differentiable at every point in the open interval (a, b).
(3) Label $f(a)$ and $f(b)$ on the vertical axis.
(4) Label the points $A(a, f(a))$ and $B(b, f(b))$. Draw a line through A and B.

b. Answer some questions.

(1) What information does the value of

$$\frac{f(b) - f(a)}{b - a}$$

give you about the line determined by the points A and B?

(2) If c is *any* point between a and b, what information does the value $f'(c)$ give you about the tangent line to the graph of f?

c. Show that the conclusion to the Mean Value Theorem for Derivatives holds.

(1) Find c between a and b so that

$$\frac{f(b) - f(a)}{b - a} = f'(c)$$

(2) Label $f(c)$ on the vertical axis. Sketch the tangent line at $C(c, f(c))$.
(3) Describe the relationship between the line connecting the points A and B and the tangent line at $x = c$.

2. Apply the Mean Value Theorem for Derivatives to some real-life situations.

a. You stop at a red light. When the light turns red, you start off slowly, continually increasing your speed. After 15 seconds you have traveled a quarter of a mile, where your motion is depicted in the following distance versus time graph:

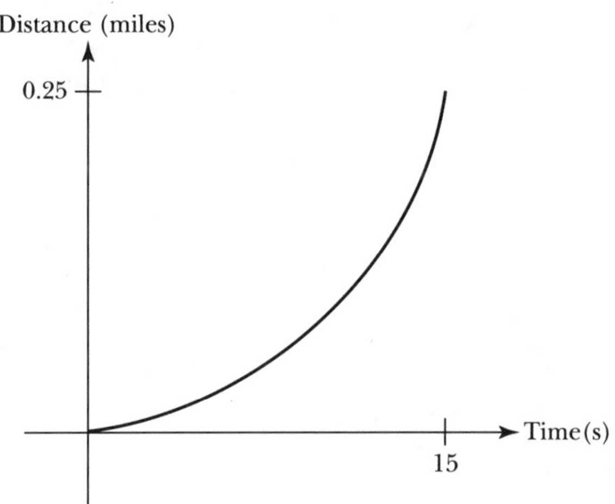

(1) Show that between 0 and 15 seconds your average speed is 60 MPH. *Hint*: 15 seconds $= \frac{1}{240}$ hours.
(2) On the graph, sketch the secant line whose slope corresponds to your average speed during the 15-second time interval.
(3) Note that your speed at any given moment corresponds to the slope of the tangent line at that point. According to the Mean Value Theorem for Derivatives, at some time c between 0 and 15 seconds your speed was 60 MPH. Label c on the t-axis. Sketch the tangent line whose slope is 60.
(4) The posted speed limit is 60 MPH. A state trooper pulls you over at 15 seconds. Do you deserve a ticket? Justify your response.

b. You work at a pretzel factory, where you make 1,200 pretzels during your 8-hour shift. The number of pretzels you can produce in t hours is given by the production function $Q(t)$, whose graph is given below:

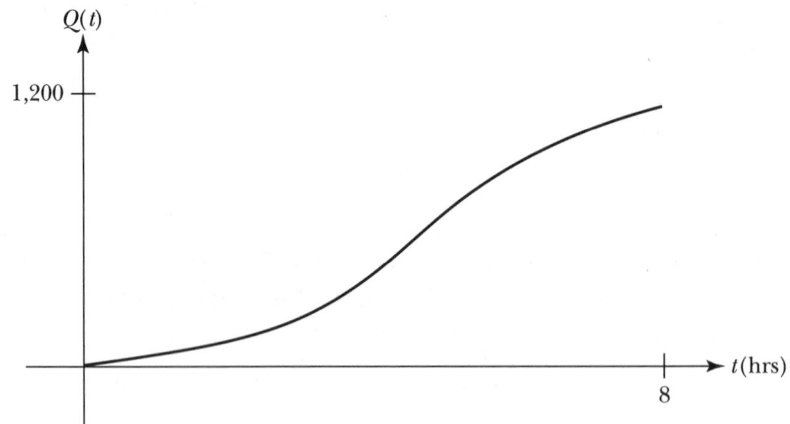

(1) Find your average rate of productivity between 0 and 8 hours.
(2) On the graph, sketch the secant line whose slope corresponds to your average rate of productivity during the 8-hour shift.
(3) According to the Mean Value Theorem for Derivatives, there exists at least one time c during your shift when your productivity rate equals your average rate of production. In your case, there are two choices for c, say, c_1 and c_2. Label c_1 and c_2 on the t-axis. Sketch the associated tangent lines.
(4) Label the time on the t-axis when your rate of productivity is the greatest. Sketch the associated tangent line.
(5) The moment of maximum efficiency, which you labeled in part (4), is called your *point of diminishing returns.* Explain why this terminology makes sense.

- Describe your understanding of derivative in HW5.8.

HW5.8 What is a derivative? Write a short essay explaining what a derivative is. In your essay:

i. List the various ways to interpret the value of a derivative.
ii. Give a formal definition of "derivative."
iii. Explain why $f'(a)$ gives the slope of the tangent line to the graph of f at $P(a,f(a))$. Support your explanation with a diagram. Label your diagram.
iv. Describe conditions under which the derivative does not exist.
v. Summarize how you can use the derivative to gain information about the behavior of its underlying function.

- Explore the use of sigma notation in HW5.9. You will use this notation in the next section.

HW5.9 Many times throughout your study of mathematics, you will be asked to find the sum of a sequence of numbers. For example, suppose your credit card company has a strict policy against late payments. For each day your payment is late, that many dollars will be added onto your total bill in addition to the amount added on for the previous days. In other words, if you are three days late, \$6 will be added onto your total bill—that is, \$3 for the third day plus \$2 for the second day and \$1 for the first day.

When there are only three terms to be added together, it is easy to write out $1 + 2 + 3$, but what if you are 100 days late? It's tedious to write out $1 + 2 + 3 + 4 + 5 + \cdots + 100$. Consequently, a special notation called *sigma notation* has been developed to represent sums of numbers. Using sigma notation, you can represent

$$1 + 2 + 3 + 4 + 5 + \cdots + 100 \text{ as } \sum_{i=1}^{100} i$$

5

The capital Greek letter sigma Σ represents a sum and the variable i is the *index* of summation. In this case, the values of i range through the integers from 1 to 100. The limits on the index will vary depending on your problem, as will the expression to be summed. For instance,

$$\sum_{i=12}^{16} i^2 \text{ represents } 12^2 + 13^2 + 14^2 + 15^2 + 16^2.$$

1. Rewrite each summation as the sum of a sequence of numbers. Do not calculate the actual value of the sum.

a. $\sum_{i=3}^{7} (2i+1)$

b. $\sum_{i=1}^{4} i^3$

c. $\sum_{i=-3}^{2} (-i+3)$

d. $\sum_{i=0}^{5} \frac{i}{i+1}$

2. Represent the following sums using sigma notation.

a. $2 + 3 + 4 + \cdots + 12$

b. $1 + 4 + 9 + 16 + \cdots + 81$

c. $1 + \frac{1}{2} + \frac{1}{3} + \frac{1}{4} + \frac{1}{5} + \cdots + \frac{1}{12}$

d. $15 + 20 + 25 + \cdots + 65$

3. Because calculating these long sums can be very tedious, it is helpful to let the computer do it for you. In ISETL, the %+ symbol is used to represent Σ. For example, to find the value of

$$48 + 75 + 108 + \cdots + 300$$

or

$$\sum_{i=4}^{10} 3i^2$$

you can enter in ISETL

```
%+ [3*i**2 : i in [4..10]];
```

where

%+ tells ISETL to sum a sequence of values
3**i***2 represents the values to be summed
i is the index of summation
4 and 10 are the lower and upper bounds on i

Look again at the sums you considered in parts 1 and 2. In each case,

i. Translate the mathematical notation for the sum into ISETL.

ii. Use ISETL to find the value of the sum.

Print a copy of your ISETL session. Label your results.

SECTION 2

The Definite Integral

The definite integral is one of the most important and fundamental concepts in calculus. It is extremely useful for measuring quantities and determining how quantities accumulate. For example, definite integrals can be used to help a maintenance worker (who has taken calculus!) figure out how much rubber sealer he will need to coat the bottom of a kidney shaped swimming pool; or the president of Chemical Bank can use a definite integral to determine his accumulated net assets on a particular day, based on the rate at which money has been deposited or withdrawn from the bank up to that time. Surprising as it may seem, these two types of calculations—finding the area of an oddly shaped region and determining an accumulation resulting from a variable rate of change—are very closely connected mathematically.

In order to be able to identify situations where it is appropriate to use a definite integral, you need to develop a conceptual understanding of what a definite integral represents. This is the primary objective of this section. The tasks have been designed to help you discover its definition and develop a mental image associated with the concept of definite integral.

In this section, you will examine the relationship between definite integrals and areas. You will use ISETL to do some of the calculations and to help you understand the mathematical notation underlying the definition of definite integral. You will explore ways to approximate areas and accumulations and think about how to improve your approximations, making them more exact. You will develop a general method which applies in all cases. Finally, you will make a first pass at examining the connection between definite integrals and derivatives.

Although this initial encounter with the concept of a definite integral will involve using a definite integral to represent the area or a region, it can be used to represent numerous quantities that accumulate. In Volume II, you will return to your study of integrals and explore other situations where the idea is applicable. In addition, you will develop a straightforward way to find exact solutions to these types of problems (without evaluating a limit).

Your study of definite integrals begins as you approximate some areas in the next three tasks.

Task 5-7: Finding Some Areas

Should your client buy this property? You have graduated from college after distinguishing yourself in this course and are working for a team of real estate consultants. One of your clients is considering buying some vacation property that borders on the Susquehanna River. The property comes equipped with a boat house and dock, but there are a few problems with them. The dock needs to be rebuilt, the back of the boat house that faces the river is in desperate need of paint, and the survey information has been lost. Your client needs to know the areas of the dock and the back of the boathouse (so she can calculate the cost of refurbishing them), and she needs to determine the area of the property (so she can determine if she can afford to buy it).

1. The dock has good supports, but the wood on the deck part is rotten. The dock deck is 4 feet wide and 6 feet long. Do a scale drawing of the deck and find its total area.

2. Suppose the local lumber mill sells leftover treated outdoor boards which are exactly 4" wide, 2" deep, and 4' long. How many boards will be needed to rebuild the deck? Sketch the boards needed on your drawing above.

3. If the treated outdoor boards cost $9.50 each and your labor is free, how much will it cost to refurbish the deck on the dock?

4. The back of the boat house is depicted in the diagram below. How many square feet need to be painted? Explain how you figured this out.

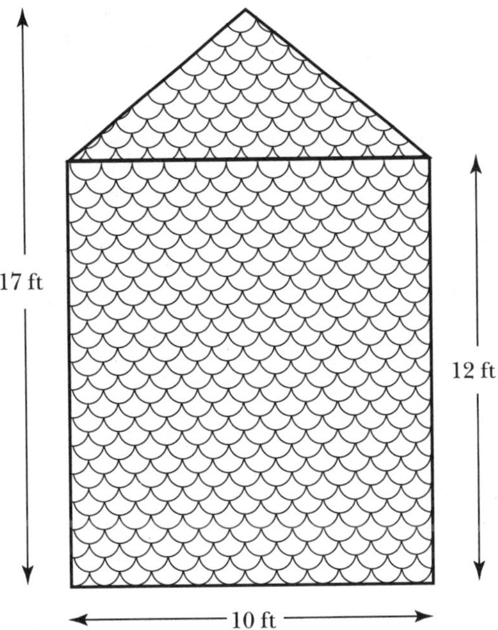

5. If a quart of special preservative costs $12.25 and covers 50 square feet. How much will the preservative cost to refurbish the back face of the boat house?

6. The seller has made a careful sketch on $\frac{1}{4}$" graph paper of the site boundaries as shown below. The property is bounded on the north by the Susquehanna River and on the south by Rt. 70. Approximately how many square feet are included in the property? Describe briefly the techniques you thought about using to find the answer, and then explain why you chose the one you used.

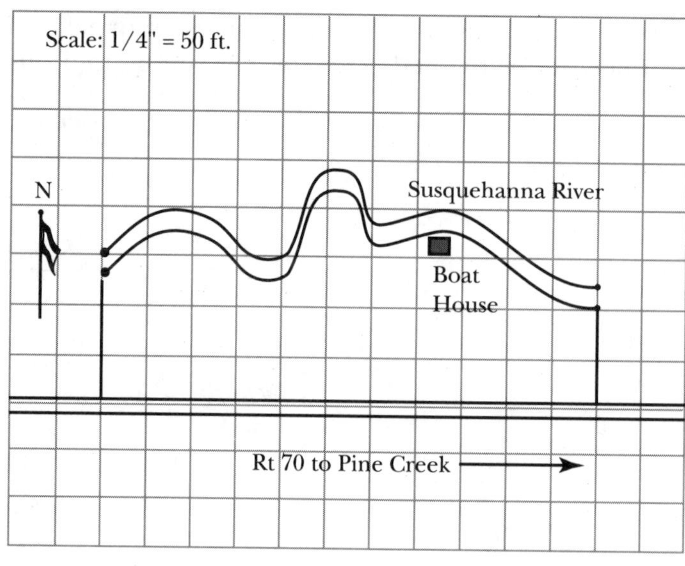

7. If the property is being sold for 1 penny per square foot, approximately how much is the seller asking for this plot of land?

8. After paying your consulting fee, your client can only afford to pay a total of $1200 for the property and the lumber and paint to fix things up. Can she afford to buy it?

In part 6 of the last task, you approximated the area of the region bounded by the Susquehanna River and Route 70. Chances are that you did this by counting the number of squares in the grid covering the region and then summing the areas of the squares. Explore some other possible ways to do this in Task 5-8.

Task 5-8: Describing Some Possible Approaches

1. First, you know how to find the areas of a variety of regular shapes. List the shapes whose areas you can calculate, and give the associated area formulas.

2. Suppose you want to find the area of an oddly shaped region lying under a given curve and over a closed interval, such as in Figure 1.

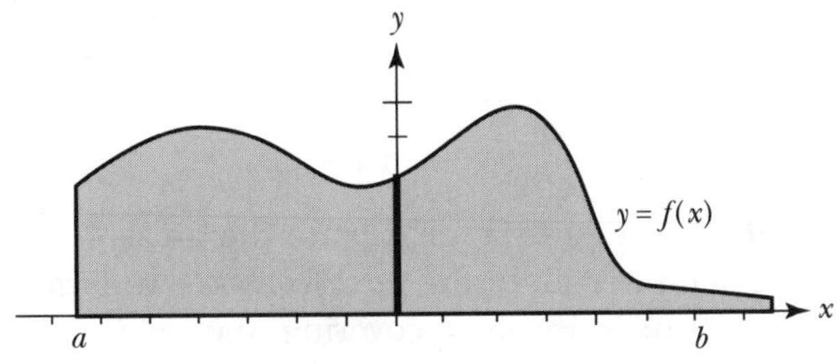

Figure 1.

For each regular shape, which you listed in part 1, make a rough sketch of the region in Figure 1. Cover the region with pieces having the desired shape. For instance, cover the region with circles. The pieces can touch, but cannot overlap. Observe that you can approximate the area of the region by summing the areas of the pieces.

There are a number of ways to approximate the area of a region under a curve and over a given interval. In the next few tasks, you will examine a rectangular approach, which involves covering the region with rectangles, formed by using the right end-point of each subinterval. To use this approach:

- Partition the given interval into equal size pieces or subintervals.
- For each subinterval, form a rectangle whose base is the subinterval and whose height is the value of the function at the right end-point of the subinterval.
- Calculate the area of each rectangle.

- Approximate the area under the curve by summing the areas of the rectangles.

Apply this approach to approximate the area of the region bounded by the Susquehanna River which you considered in Task 5-7, part 6.

Task 5-9: Applying a Rectangular Approach

Note: A blowup of the region under consideration appears on the next page. Each square in the grid is assumed to be $\frac{1}{4}$" by $\frac{1}{4}$".

1. Place the region on a coordinate system, with the southwest corner of the property at the origin and the horizontal and vertical axes determined by the highway and the western boundary respectively. Carefully label the axes in terms of feet.

2. Suppose h is the function whose value at points along the southern boundary is determined by measuring the distance from the highway to the riverbank. For example, if you look carefully at the diagram, $h(0)$ is approximately 131 feet. Approximate the value of h at each of the following points along the southern boundary:

$$h(0) = 131'$$

$$h(25) =$$

$$h(100) =$$

$$h(275) =$$

$$h(350) =$$

3. Suppose you wanted to approximate of the area of the property by dividing the property into 11 rectangular strips of equal widths, running north to south. A mathematician refers to the base of each strip as a *subinterval.* Find the size of each subinterval.

4. Assume the height of each rectangular strip is determined by the value of h at the *right* end-point of the associated subinterval.

a. Sketch the 11 rectangles on the blowup of the region.

5

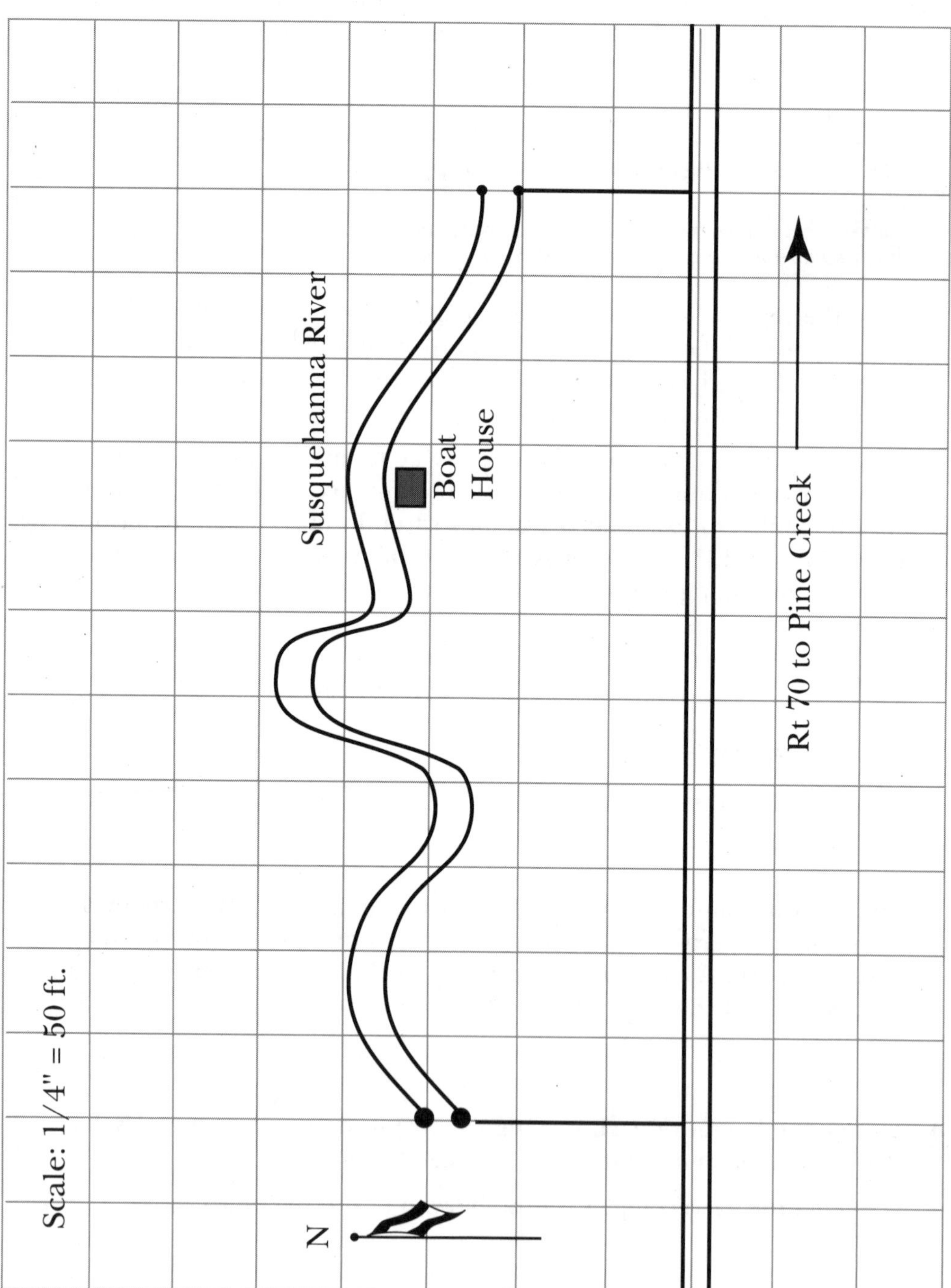
Susquehanna River
Boat
House
Rt 70 to Pine Creek
Scale: 1/4" = 50 ft.
N

b. First make sure that the column headings on the chart given below make sense to you, then fill in the chart.

Subinterval # i	Right End-Point of ith Subinterval x(ft)	Height of ith Rectangle $h(x)$ (ft)	Width of ith Rectangle Δx (ft)	Area of ith Rectangle $h(x) \cdot \Delta x$ (ft^2)
1	50			
2	100			
3	150			
4	200			
5	250			
6	300			
7	350			
8	400			
9	450			
10	500			
11	550			

c. Use the information on the chart to approximate the area of the region. How does this result compare to the approximation you found in Task 5-7, part 6?

d. Describe how you can make your approximation more precise.

In the last task, you approximated the area of a region by using a small number of rectangles. Your goal is to find the actual value. Intuitively, it should make sense that the more rectangles you consider, the more accurate your approximation will be. In the next task, you will develop the mathematical notation for the general situation. You will represent the approximate area of a region by an expression and then pass to the limit to represent the exact area.

Task 5-10: Considering the General Situation

Suppose f is a non-negative, continuous function, where $a \leq x \leq b$. Consider the region under the graph of f and over the closed interval $[a,b]$. Approximate the area of this region by covering it with n rectangles and then summing the areas of the rectangles. Express the exact area in terms of a limit.

1. Sketch the graph of f, for $a \leq x \leq b$. Make the graph of f curvaceous and make it lie above the x-axis.

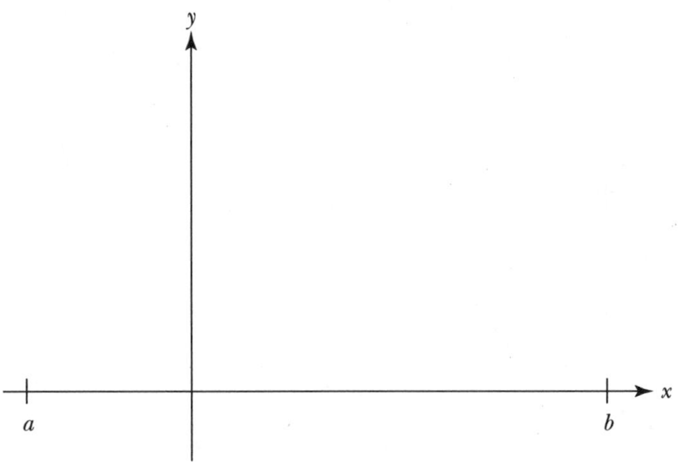

2. Partition the closed interval $[a,b]$ into n equal subintervals and find a formula for the right end-point of each subinterval. These subintervals will form the bases of the rectangles that you will use to approximate the area of the region.

   a. Observe that the length of the entire interval is $b - a$. Find a formula for the width of each of the n subintervals in terms of a, b, and n. Call this width Δx—which is read "delta x" and denotes a small change in x.

   $$\Delta x =$$

   b. Find formulas for the end-points of the n subintervals and give each end-point a name.

   The initial end-point is a. Call this x_0; that is,

   $$x_0 = a$$

(1) The right end-point of the first subinterval is Δx away from a. Find a formula for the end-point of this subinterval in terms of a and Δx. Call this end-point x_1.

$x_1 =$

(2) The end-points of the successive subintervals are Δx apart. Find formulas for the right end-points of the second and third subintervals in terms of a and a multiple of Δx. Call the end-points of these subintervals x_2 and x_3, respectively.

$x_2 =$

$x_3 =$

(3) Recall that n represents the total number of subintervals. You know that n is a positive integer, but you don't know which one. Consequently, it is impossible to write out all the values of the right end-points of the subintervals. What might you do in this case? One approach is to consider an arbitrary subinterval—that is, instead of referring to the first, second, or third subinterval, refer to the ith subinterval where $1 \leq i \leq n$, as there are n subintervals.

Use the pattern that emerged when you found formulas for x_1, x_2, and x_3 to find a formula for the right end-point of the ith subinterval. As before, express the formula in terms of a and a multiple of Δx. Call the end-point of the ith subinterval x_i.

$x_i =$, where $1 \leq i \leq n$

(4) The right end-point of the last, or nth, subinterval is b. Find a formula for the right end-point of this subinterval in terms of a and a multiple of Δx.

$x_n = b =$

c. Mark the values of x_i, where $0 \leq i \leq n$, on the x-axis on your diagram in part 1.

(1) Relabel a as x_0 and b as x_n.

(2) Starting at the value for a, make three, equally spaced tick marks on the x-axis, representing the right end-points of the first three subintervals. Label them x_1, x_2, and x_3.

(3) Because it's impossible to label the end-point of every subinterval, after the tick mark for x_3, put three dots on the x-axis to indicate that some points are missing.

(4) Make tick marks for x_{i-1} and x_i, which represent the left and right end-points of the ith subinterval, and label them appropriately.

(5) After the tick mark for x_i, put three more dots on the x-axis to indicate that more points are missing.

(6) Finally, noting that b is x_n, make a tick mark for x_{n-1}, which is the left end-point of the last—or nth—subinterval.

3. Cover the region with rectangles and then approximate the area under the graph of f over $[a,b]$ by summing the areas of these rectangles.

a. Sketch a few rectangles. On your diagram in part 1, sketch the first three rectangles, the ith rectangle, and the nth rectangle, using the value f at the right end-point of each subinterval to determine each rectangle's height.

b. Find a formula for the height of each rectangle. Express the height in terms of f and the right end-point of the subinterval.

(1) Express the height of the first rectangle in terms of f and x_1.

(2) Express the height of the second rectangle in terms of f and x_2.

(3) Express the height of the ith rectangle in terms of f and x_i.

(4) Express the height of the nth rectangle in terms of f and x_n.

c. Find a formula for the area of each rectangle. Recall that the width of each rectangle is Δx.

(1) Find a formula for the area of the first rectangle.

(2) Find a formula for the area of the second rectangle.

(3) Find a formula for the area of the ith rectangle.

(4) Find a formula for the area of the nth rectangle.

4. Approximate the area of the region by summing the areas of the rectangles. Express the sum using sigma notation. This sum is called a *Riemann sum.*

5. What happens to the value of the Riemann sum if the number of rectangles is allowed to increase indefinitely? How might you express this using limit notation?

Congratulations? You have just developed the notation for what mathematicians call the *definite integral,* another extremely important calculus concept.

In the last task, you took a concrete idea—adding up the areas of rectangles—and represented it in an abstract way using mathematical notation. In particular, you considered a continuous, non-negative function f defined over a closed interval $[a,b]$. You let

$$\Delta x = \frac{b-a}{n} \qquad \text{and} \qquad x_i = a + i\Delta x \quad \text{for } 0 \leq i \leq n$$

where n is a positive integer.

5

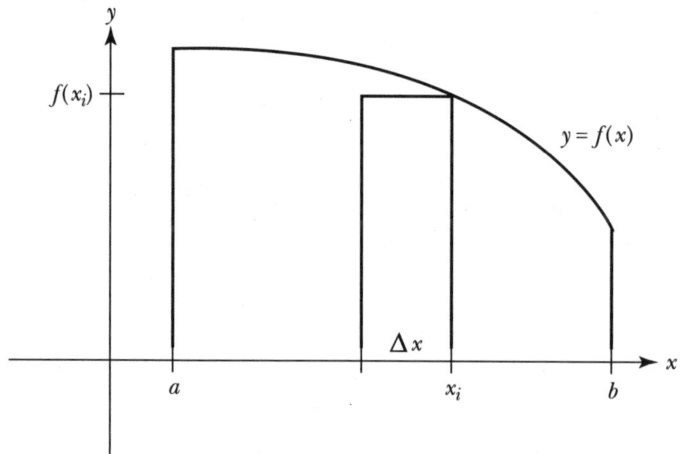

You discovered that

$$\underbrace{\lim_{n \to \infty} \underbrace{\sum_{i=1}^{n} \underbrace{f(x_i)\Delta x}_{\text{area of } i\text{th rectangle}}}_{\text{approximate area of region under curve}}}_{\text{exact area of region under the graph of } f \text{ over the interval } [a,b]}$$

The approximate area is represented by a Riemann sum. In Task 5-9, you calculated a Riemann sum by hand. Obviously, when you do a calculation by hand, you need to keep the number of rectangles—that is, value of n—very small. Larger values of n, however, give more precise approximations. This sounds like a perfect task to do on a calculator or with the computer. In the next task, you will use ISETL to calculate some Riemann sums and observe the limiting behavior of the Riemann sums as the size of n increases. Using ISETL will not only help you do the calculations, it should also help you develop a better understanding of the mathematical notation.

Task 5-11: Calculating Riemann Sums

1. Approximate the area of the region bounded by the x-axis and the graph of

$$f(x) = x^2 - 6x + 10, \quad \text{where } 2 \le x \le 5$$

a. Consider the following graph of f.

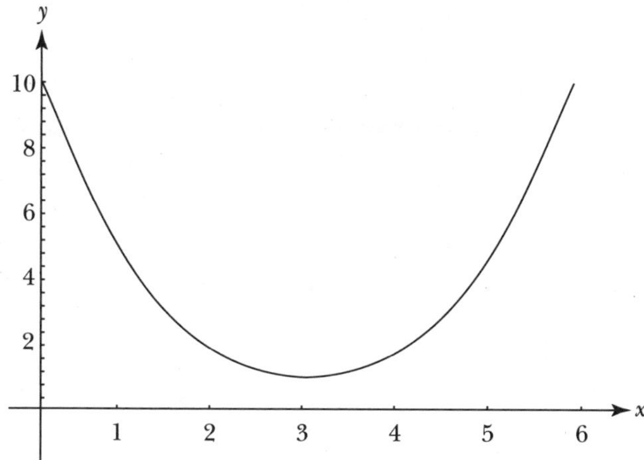

(1) Shade the area of the region under the graph of f and over the closed interval [2,5].

(2) Suppose you approximate the area of this region using six rectangles, which all have the same width and where the height of each rectangle is determined by the value of f at the right endpoint of its base. Draw the six rectangles, using a pencil and a straightedge.

(3) If you use six rectangles to estimate the area, will the approximation be greater than, equal to, or less than the exact value? Justify your response.

(4) Suppose you use 10 rectangles, or 100, or 1000 to estimate the area. How will the approximate value compare to the exact value?

b. Use ISETL to approximate the area of the region under the graph of $f(x) = x^2 - 6x + 10$ and over the closed interval [2,5], using six rectangles.

5

(1) Enter the following code in ISETL. Try to connect the ISETL syntax with the associated mathematical notation which you developed in Task 5-10, noting that in this example, $a = 2$, $b = 5$, and $n = 6$.

ISETL Code

```
$ Define the function f(x) = x^2 - 6x + 10.

f := func(x);
        return x**2 - 6*x + 10;
end func;

$ Assign values to a, b, n.

a := 2;
b := 5;
n := 6;

$ Define Δx.

delta_x := (b - a)/n;
```

```
$ Define the right end-points of the rectangles x_i = a + iΔx where 1 ≤ i ≤ n.

x := [a + i * delta_x : i in [1..n]];

$ Note: x is a sequence containing n values, whose first item x(1) corresponds to
x_1, second item x(2) corresponds to x_2, and so on. To display the entire sequence,
enter x;. To access a particular item, for instance the fifth one, enter x(5);.
```

```
$ Calculate the Riemann sum Σ_{i=1}^{n} f(x_i)Δx. (See HW5.9.)

%+ [f(x(i)) * delta_x : i in [1..n]];
```

```
$ Construct a table containing the area of each of the n rectangles.

for i in [1..n] do
        writeln i, x(i), f(x(i)), f(x(i)) * delta_x;
end for;

$ Note: This table is similar to the one you constructed by hand in Task 5-9, part 4.
```

(2) Print a copy of the ISETL table and place it in the space below. Label the columns. Explain how the information in the table relates to the diagram you drew in part a.

(3) Based on your ISETL session, what is the approximate area of the region when you use a rectangular approach with the right end-point of each subinterval, where $n = 6$?

c. Use ISETL to make your approximation more precise.

Note: To calculate the Riemann sum for a new value of n:

i. Edit and reexecute the assignment statement for n.

ii. Reexecute the assignment statements for **delta_x** and **x** and the expression for the Riemann sum, which all depend on the value of n.

(1) Use ISETL to fill in the following chart.

n	$\sum_{i=1}^{n} f(x_i)\Delta x$
5	
15	
25	
40	
75	
100	

(2) As n gets larger and larger, the Riemann sum gets closer and closer to the exact value of the area. Based on the chart, what might you conclude about the limiting behavior of the associated Riemann sum?

$$\sum_{i=1}^{n} f(x_i)\Delta x \rightarrow \boxed{} \quad \text{as } n \rightarrow \infty$$

2. Compare the mathematical notation developed in Task 5-10 to the associated ISETL syntax by filling in the boxes in the table given below.

Note: A series of three dots (...) indicates that if you were considering a specific situation, you would insert more information here. For example, you would define f in the first row. Do not fill in the ...'s. Think in terms of a general situation.

Interpretation	Math Notation	ISETL Syntax
Function definition	$f(x) = \ldots$	f := func(x); end func;
Left end-point of interval	$a = \ldots$	
Right end-point of interval		b := ...;
# of rectangles		
Width of ith rectangle		delta_x := (b − a)/n;
Right end-points of ith subinterval	$x_i = a + i\Delta x,\ 1 \le i \le n$	
Riemann Sum with n rectangles		%+[f(x(i)) * delta_x : i in [1..n]];

3. Consider the region bounded by the graph of $g(x) = e^x$ and the x-axis, where $-1 \leq x \leq 1$.

a. Use your CAS to plot the region.

(1) Place a copy of the graph in the space below.

(2) On the graph, shade the region under consideration.

(3) Suppose $n = 8$. Sketch the eight rectangles you would use to approximate the area of the region using the rectangular approach with the right end-point of each subinterval.

b. Use ISETL to calculate the value of the Riemann sum for $n = 8$. Enter your result below.

c. Explain why using the right end-point of each subinterval to determine the height of the associated rectangle will always result in an approximation that is larger than the exact value.

d. The area of the region under the graph of $g(x) = e^x$ and over the closed interval $-1 \leq x \leq 1$ is very close to 2.3504. Use ISETL to determine the minimum number of rectangles you need to consider so that the difference between the Riemann sum and 2.3504 is less than or equal to 0.05. Record the results of your trials in the table below.

Trial Number	n	$\sum_{i=1}^{n} f(x_i)\Delta x$
1		
2		
3		

If the limit of the Riemann sum $\sum_{i=1}^{n} f(x_i)\Delta x$ exists as $n \to \infty$, f is said to be *integrable* over the closed interval $[a,b]$. A shortcut for denoting this limit is to use the definite integral notation

$$\int_a^b f(x)\,dx$$

which is read, "the integral from a to b of $f(x)\ dx$."

You can express a definite integral two ways: using the expression representing a function or using the name of a function. For example, in the last task, you considered the region bounded by

$$f(x) = x^2 - 6x + 10, \quad \text{where } 2 \le x \le 5$$

In this case, because you know the expression, $x^2 - 6x + 10$, as well as the name, f, you can represent the area under the graph of f over the closed interval $[2,5]$ by

$$\int_2^5 (x^2 - 6x + 10)\,dx \qquad \text{or} \qquad \int_2^5 f(x)\,dx$$

Evaluating a definite integral gives a number, such as 2.3504, −260.5, or 0. For instance, because the Riemann sums approximating the area under the graph of $f(x) = x^2 - 6x + 10$ over the interval [2,5] approach 6 as $n \to \infty$, you write

$$\int_2^5 (x^2 - 6x + 10)\, dx = 6$$

If the region under the graph of a function and over a closed interval has a regular shape, you can find the exact value of the associated definite integral by using the area formula for the regular shape. However, based on the work you have done so far, if the region is oddly shaped, you can only approximate the value of the definite integral by calculating a Riemann sum. The big question, of course, is how you might evaluate a definite integral without taking the limit of a Riemann sum.

The next task will help you to get used to using integral notation and to think about interpreting the notation as an area. You will convert back and forth between the description of a region and its associated definite integral. You will use a geometric approach to evaluate definite integrals that correspond to regularly shaped regions.

Task 5-12: Interpreting Definite Integrals

1. Consider the region bounded by $f(x) = x - 4$ over the closed interval [6,10].

a. Sketch the function and shade the designated region.

b. Represent the area of the region by a definite integral.

c. Find the value of the definite integral by using a geometric formula to calculate the area of the region.

2. Represent the area of each of the following regions by a definite integral.

a. The region under the graph of $g(x) = x^4 + 3x + 8$, where $-5 \leq x \leq 10$.

b. The region bounded by $x = -2$, $x = 5$, $y = 0$, and $y = 8$.

c. The shaded region given below.

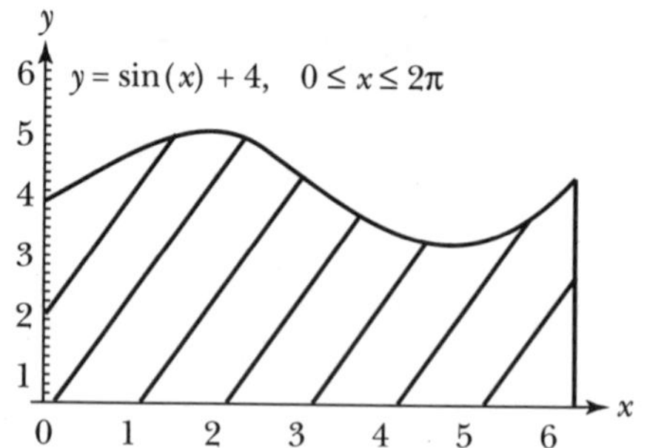

d. The shaded region given below.

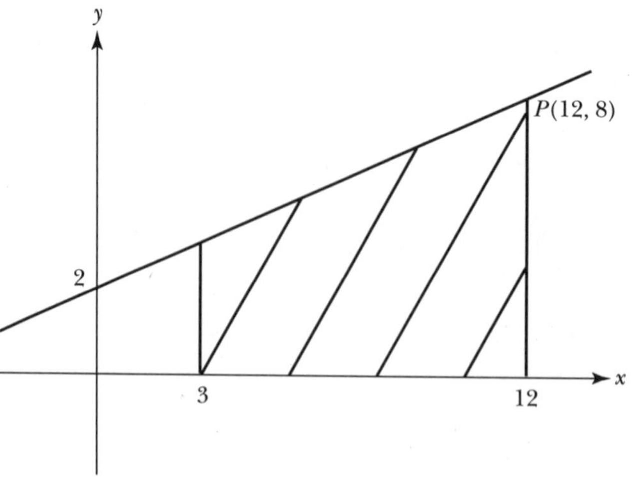

3. Sketch the region whose area is represented by each of the following definite integrals.

a. $\int_{-2}^{3} x^2\,dx$

b. $\int_{-\frac{\pi}{2}}^{\frac{\pi}{2}} \cos(t)\,dt$

c. $\int_{-2}^{0} (x+3)\,dx + \int_{0}^{2} (-x+3)\,dx$

4. Use a geometric approach to evaluate the following definite integrals.

Note: Before evaluating the integral, sketch the associated region.

a. Evaluate $\int_{0}^{6} x\,dx$.

b. Evaluate $\int_{-1}^{2} (2x+6)\,dx$.

Note: The area of a trapezoid is given by $\frac{1}{2}(h_1 + h_2)b$, where h_1 and h_2 are the heights of the two sides and b is the width of the base.

5

c. Evaluate $\int_{-2}^{3.5} dx$.

d. Evaluate $\int_{-3}^{1} |x|\, dx$.

e. Evaluate $\int_{1}^{7} g(x)\, dx$, where g is a semicircle with radius 3 and center at $C(4,0)$.

f. Evaluate $\int_{-10}^{-5} 3.5\, dx$.

5. Suppose f is a non-negative, continuous function, defined at $x = a$. Use an area-based argument to explain why $\int_{a}^{a} f(x)\, dx$ is equal to 0.

In this unit, you have developed definitions for the two most important concepts in calculus: the derivative and the definite integral. At first glance, these concepts seem totally unrelated—one measures rates of change, whereas the other measures the size of a quantity. It turns out that they are related and that this relationship will provide you with a straightforward way (which does not involve calculating a limit) to evaluate definite integrals. In the next task, you will take a first look at the connection between these two fundamental ideas. Discovering how this connection can help you evaluate definite integrals will have to wait until Volume II.

Before investigating the connection, think about how integrals can be used to measure quantities that accumulate. Consider, for example, a non-negative, continuous function f defined for $1 \leq x \leq 10$. You can find the area of the region under the graph of f and over the closed interval $[1,10]$ by covering the region with rectangles, calculating the corresponding Riemann sum, and then taking the limit. You can express this result in terms of a definite integral:

$$\int_1^{10} f(t)\,dt = \text{area under the graph of } f \text{ and over the closed interval } [1,10]$$

This definite integral gives the value of the area of the entire region. If, however, you only wanted to calculate the area of the region under the graph of f and over the closed interval $[1,6.5]$, you could *truncate* the Riemann sum you used to calculate the entire area—that is, delete the rectangles corresponding to $[6.5,10]$—and then take the limit of this *partial Riemann sum.* You can express this new result as a definite integral:

$$\int_1^{6.5} f(t)\,dt = \text{area under the graph of } f \text{ and over the closed interval } [1,6.5]$$

In fact, you can find the area under f and over the closed interval $[1,x]$, where x is *any* value between 1 and 10, by taking the limit of the corresponding partial Riemann sum. As above, you can express this result as a definite integral:

$$\int_1^{x} f(t)\,dt = \text{area under the graph of } f \text{ and over the closed interval } [1,x], \quad \text{where } 1 \leq x \leq 10$$

Because for each value of x between 1 and 10 the value of the definite integral gives the area of the corresponding region, this process defines a new function, which we will call A:

$$A(x) = \int_1^{x} f(t)\,dt, \quad \text{where } 1 \leq x \leq 10$$

A is said to be the *accumulation function defined by f,* or in this case, the *cumulative area function,* because whenever you give A an x-value between 1 and 10, A gives the accumulated area under the graph of f from 1 to x.

Now comes the connection between derivatives and definite integrals. According to the *Fundamental Theorem of Calculus,* the derivative of the accumulation function is equal to its underlying function, or $A'(x) = f(x)$.

5

Consequently, you integrate f to define A and differentiate A to get f. In the next task, you will show that this connection holds for a constant function.

Task 5-13: Checking the Connection Between Derivatives and Definite Integrals

Consider the function $f(x) = 5$, where $2 \leq x \leq 9$. Define the associated cumulative area function and show that its derivative equals 5.

1. Sketch a graph f for $2 \leq x \leq 9$.

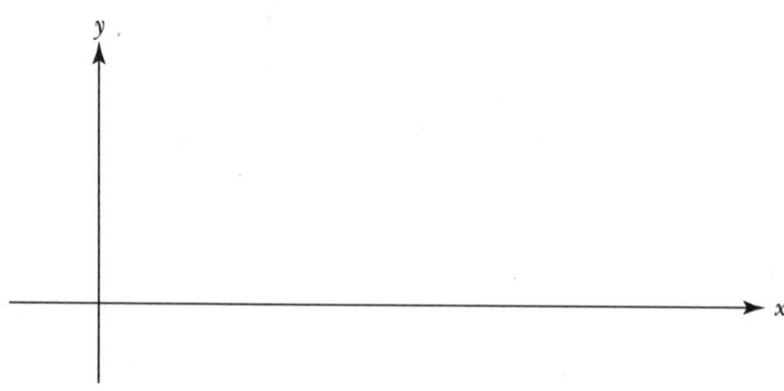

2. Find the cumulative area function A defined by f.

Note that in this case, $A(x)$ equals the area of the region under the graph of f and over the closed interval $[2,x]$, where x is any value between 2 and 9 inclusive.

a. Interpret some values of $A(x)$.

(1) For each of the following, make a small sketch of f and shade the associated region.

(a) $A(4)$

(b) $A(6.5)$

(c) $A(9)$

(2) Explain why $A(2) = 0$.

(3) Give the domain of A.

b. Use the graph of f to calculate some numerical values of $A(x)$. Represent each value as a definite integral.

	x				
	2	3	7.5	8	9
$A(x)$ (numerical value)					
$A(x)$ (represented by a definite integral)					

c. Sketch a graph A for $2 \leq x \leq 9$.

5

d. Represent A by an expression.

3. You now have *two* functions, the original function $f(x) = 5$ and a new function $A(x) = 5x$. A is the accumulation function associated with f, which is defined by integrating f. The Fundamental Theorem of Calculus claims that the derivative of the accumulation function equals the original function—that is, $A'(x) = f(x)$. So, in this instance, $A'(x)$ should equal 5. Show that this is true, using both a geometric approach and the definition of derivative.

a. Use a graphic approach to find $A'(x)$.

Consider the graph of $A(x) = 5x$. Show that the slope of the tangent line to any point on the graph of A is 5.

b. Use the definition of derivative to find $A'(x)$.

Show that

$$\lim_{h \to 0} \frac{A(x+h) - A(x)}{h} = 5, \quad \text{when } A(x) = 5x.$$

Unit 5 Homework After Section 2

- Complete the tasks in Section Two in your Activity Guide. Be prepared to discuss them in class.
- In Task 5-11 you estimated the area of a given region using a rectangular approach, where the height of each rectangle was determined by the value of the function at the right end-point of the associated subinterval. Consider three other approaches for estimating this area in HW5.10.

HW5.10 Consider, once again, the region bounded by the x-axis and the graph of

$$f(x) = x^2 - 6x + 10, \quad \text{where } 2 \leq x \leq 5$$

1. In parts a–c below, estimate the area of the region under the graph of f and over the closed intervals [2,5] using the indicated approach. In each case:

   i. Sketch the graph of f for $2 \leq x \leq 5$.
   ii. Subdivide the interval [2,5] into six equal subintervals. Cover the region with the specified shapes. For example, in part a, cover the region with rectangles formed by using the value of f at the left end-point of each subinterval.
   iii. Approximate the area of the region by finding the areas of the specified shapes and summing the results.

   a. Rectangular approach using left end-point: Form rectangles where the height of each rectangle is determined by the value of f at the left end-point of each subinterval.

   b. Midpoint approach: Form rectangles where the height of each rectangle is determined by the value of f at the midpoint of each subinterval.

   c. Trapezoidal approach: Form trapezoids where the heights of each trapezoid is determined by the value of f at the left and right end-points of each subinterval.

2. Which approach (rectangular approach using the right end-point, rectangular approach using the left end-point, midpoint approach, or trapezoidal approach) appears to give the most accurate result for this problem. Why?

- Investigate how the value of an approximation relates to the exact value when using different approaches.

HW5.11 Suppose you want to approximate the area of a region under a given curve and over a closed interval. For each of the following statements:

i. Sketch a graph of a function for which the statement is true.
ii. Shade the region.
iii. Sketch a typical rectangle or trapezoid. (Your example should work for any number of rectangles or trapezoids.)

1. You approximate the area using the rectangular approach with the left end-point of each subinterval. The approximate area is greater than the exact area.

2. You approximate the area using the rectangular approach with the left end-point of each subinterval. The approximate area is less than the exact area.

3. You approximate the area using the rectangular approach with the left end-point of each subinterval. The approximate area equals the exact area.

4. You approximate the area using the rectangular approach with the right end-point of each subinterval. The approximate area is greater than the exact area.

5. You approximate the area using the rectangular approach with the right end-point of each subinterval. The approximate area is less than the exact area.

6. You approximate the area using the rectangular approach with the right end-point of each subinterval. The approximate area equals the exact area.

7. You approximate the area using the trapezoidal approach. The approximate area is greater than the exact area.

8. You approximate the area using the trapezoidal approach. The approximate area is less than the exact area.

9. You approximate the area using the trapezoidal approach. The approximate area equals the exact area.

- Model some situations and interpret some definite integrals in HW5.12.

HW5.12

1. Model some familiar situations and represent the regions by definite integrals.

Look again at the dock and the boat house which you considered in Task 5-7, parts 1 and 4. In each case:

i. Model the situation by placing the region on a coordinate system and finding an expression for the function which defines the shape of the region.
ii. Express the exact area of the region in terms of a definite integral using the expression from part i.
iii. Evaluate the integral using a geometric approach.

a. The deck of the dock.

b. The back of the boat house. (*Note:* The upper boundary of the back of the boat house is modeled by two expressions. Consequently, its area can be represented by the sum of two definite integrals. It is also possible to represent the area as a multiple of a single definite integral. Try to do it both ways.)

2. Consider

$$h(r) = \begin{cases} r+1, & \text{if } r \le 3 \\ 4, & \text{if } 3 < r < 5 \\ -4r+24, & \text{if } r \ge 5 \end{cases}$$

Use a geometric approach to evaluate $\int_0^6 h(r)\,dr.$

3. Sketch the region whose area is given by $\int_0^2 x^2\,dx + \int_2^4 (x-4)^2\,dx$.

4. Suppose f is a non-negative continuous function with domain $1 \le x \le 3$, where the area of the region under the graph of f over the closed interval [1,3] is greater than or equal to $\frac{1}{2}f(1.5) + \frac{1}{2}f(2) + \frac{1}{2}f(2.5) + \frac{1}{2}f(3)$. Sketch a graph of f.

5. Sketch the graph of a function g, where $\int_{-4}^{-1} g(x)\,dx = 15/2$.

6. Explain why the following inequalities are true.

a. $\int_0^{\pi} \sin(\theta)\,d\theta < \pi$

b. $\int_0^4 f(t)\,dt < 24$, where

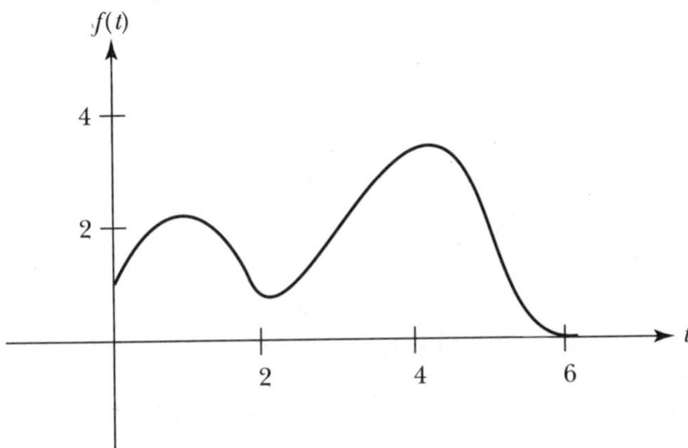

• Show that the Fundamental Theorem of Calculus holds for a linear function.

HW5.13 Consider $f(x) = 2x$, where $0 \le x \le 4$.

1. Graph f for $0 \le x \le 4$.

2. Let A be the accumulation function defined by f.

a. Interpret some values of $A(x)$. For each of the following values of x:

i. Draw a small sketch of f and shade the region whose area equals $A(x)$.

ii. Find the numerical value of $A(x)$.

5

iii. Represent $A(x)$ by a definite integral.

(1) $x = 0$	**(4)** $x = 2.5$
(2) $x = 1$	**(5)** $x = 3$
(3) $x = 2$	**(6)** $x = 4$

b. Sketch a graph of A for $0 \leq x \leq 4$.

c. Find an expression that fits the graph of A.

3. Show that the Fundamental Theorem of Calculus holds. Use the definition of the derivative to show that $A'(x) = f(x)$.

- Describe your understanding of definite integral in HW5.14.

HW5.14 What is a definite integral? Write a short essay explaining what a definite integral is. In your essay:

i. Give the definition of a definite integral as the limit of a Riemann sum.
ii. Support your definition by sketching and labeling an appropriate diagram.

- Write your journal entry for this unit. As usual, before you begin to write, review the material in the unit. Think about how it all fits together. Try to identify what, if anything, is still causing you trouble.

HW5.15 Write your journal entry for Unit 5.

1. Reflect on what you have learned in this unit. Describe in your own words the concepts that you studied and what you learned about them. How do they fit together? What concepts were easy? Hard? What were the important ideas? Give some examples of the main ideas.

2. Reflect on the learning environment for the course. Describe the aspects of this unit and the learning environment that helped you understand the concepts you studied. What activities did you like? Dislike?

Index